0304D12

Technoculture and Critical Theory

While technology allows humanity to develop more constructive ways of engaging with the world, it also extends the capacity for domination and exploitation. It enables us to realise our needs, but can also reconstruct those needs so that aspirations towards human progress realise themselves within a technocratic, antihuman paradigm – in the 'service of the machine'. *Technoculture and Critical Theory* theorises the ambivalence most of us register towards technological progress.

The author explores the work of major thinkers and cultural movements that have grappled with the complex relationship between technology, politics and culture. Subjects such as the Internet, cloning, warfare, fascism and Virtual Reality are placed within a broad theoretical context which explores how humanity might, through technology, establish a more ethical relationship with the world.

Examining the philosophy of writers such as Heidegger, Benjamin, Lyotard, Virilio and Žižek, and cultural movements such as Italian Futurism, this book marks a timely intervention in critical theory debates. The broad scope of the book will be of vital interest to those in the fields of philosophy, critical theory, cultural studies, politics and communications.

Simon Cooper is an editor of *Arena Journal*, and teaches in Communications at Monash University.

Routledge Studies in Science, Technology and Society

1 Science and the Media
Alternative routes in scientific communication
Massimiano Bucchi

2 Animals, Disease and Human Society
Human–animal relations and the rise of veterinary medicine
Joanna Swabe

3 Transnational Environmental Policy
The Ozone Layer
Reiner Grundmann

4 Biology and Political Science
Robert H. Blank and Samuel M. Hines, Jr.

5 Technoculture and Critical Theory
In the service of the machine?
Simon Cooper

Technoculture and Critical Theory
In the service of the machine?

Simon Cooper

London and New York

First published 2002
by Routledge
11 New Fetter Lane, London EC4P 4EE

Simultaneously published in the USA and Canada
by Routledge
29 West 35th Street, New York, NY 10001

Routledge is an imprint of the Taylor & Francis Group

© 2002 Simon Cooper

Typeset in Garamond by
M Rules
Printed and bound in Great Britain by
Biddles Ltd, Guildford and King's Lynn

All rights reserved. No part of this book may be reprinted or
reproduced or utilised in any form or by any electronic,
mechanical, or other means, now known or hereafter
invented, including photocopying and recording, or in any
information storage or retrieval system, without permission in
writing from the publishers.

British Library Cataloguing in Publication Data
A catalogue record for this book is available from the British Library

Library of Congress Cataloguing in Publication Data
A catalog record for this book has been requested

ISBN 0–415–26160–0

Contents

Acknowledgements vii

1 Introduction: in the service of the machine? 1

2 Beyond enframing: Heidegger and the question concerning technology 18

3 Walter Benjamin and technology: social form and the recovery of aura 44

4 Futurism and the politics of a technological being in the world 67

5 Between totalitarianism and heterogeneity: Lyotard and the postmodern condition 88

6 Paul Virilio: overcoming inertia? 114

7 Psychoanalysis, cyberspace and its discontents: Turkle, Žižek, Brennan 138

8 Conclusion 160

Notes 166
Bibliography 171
Index 177

Acknowledgements

This text began as a doctoral dissertation and so I would first like to thank my supervisors Andrew Milner and Paul James for their assistance and advice in the preparation of that earlier work.

I would especially like to thank my friends and colleagues in the *Arena* group, whose theoretical approaches, friendship and encouragement have provided an important context through which my own ideas could develop. In particular I would like to thank Geoff Sharp, but also John Hinkson, Alison Caddick, Paul James, Guy Rundle and Nonie Sharp for their support, advice, and for providing a model of political, theoretical, and personal engagement that has profoundly affected my own work and life. The contribution of *Arena* to Australian political and cultural life is enormous; without it, I suspect this work could never have been written.

Long-suffering friends who deserve thanks for their friendship and support are Paul Atkinson, Andrew Johnson, Ben Rossiter, Jacqueline Coad, Alison Wall, Matthew Ryan and Mary Roberts, Lauren, Bonnie and Ian Black, and Gail Ward. Thanks also to my new colleagues within the mass communications section at Monash, in particular Mary Griffiths and Mike Griffiths.

I am especially grateful to Alison Hart for all her help and generous support in the preparation of the book manuscript, often in difficult circumstances. Her energy, intelligence and focus allowed the text to gain shape much more rapidly than it might have otherwise.

The most profound debts are the hardest to express. I want to thank my mother for all her support and encouragement, and for allowing me the freedom to pursue my own path. But most of all I want to thank Samantha Black for her love and support. She has long been a source of inspiration, intelligence and humour in my life. Without Samantha, it is unlikely that the book would have been completed. I therefore dedicate it to her.

Sections of Chapters 2, 3 and 6 have appeared in an earlier form in *Arena Journal* (see no. 5, 1995; no. 6, 1996, and no. 9, 1997).

1 Introduction

In the service of the machine?

> the problem is knowing whether the Master–Slave conflict will find its resolution in the service of the machine.
>
> (Lacan 1977: 27)

Lacan's 'problem' encapsulates the ambivalence most of us register towards technological progress. While technology allows humanity to develop more constructive ways of engaging with the world, it also extends the capacity for domination – whether it be nature or simply those who are different. Technology can allow us to realise our needs; however it can also reconstruct those needs so that aspirations towards human progress realise themselves within a technocratic, antihuman paradigm – in the 'service of the machine'. This book is about how to come to terms with this ambivalence.

Implicitly, we all recognise that technology occupies an increasingly central role in our lives. Whether we look at the Internet, the possibilities for human cloning, the rise of high-tech global markets or any number of other examples, there seems to be scarcely any part of our lives that is not in some way technologically mediated. Popular culture registers this phenomenon of an increased technologisation of the lifeworld by simultaneously welcoming the change and contradictorily creating narratives revolving around a peculiar 'paranoid' sensibility. Cyberpunk fiction, television shows like the *X-Files* and *Nowhere Man*, and movies such as *The Matrix* tap into our fears of the increased capacity of technology to affect our lives, whether through more pervasive surveillance mechanisms or through the manufacture of powerful technological illusions. At one level, the increased *popularity* of paranoia as a cultural fantasia registers our ambivalence towards the effects of this ever-increasing technological mediation of our lives. At a deeper level, paranoia might be an implicit recognition of how technology works subtly, and behind our backs, to reconstruct the mode of our being human. This is the central concern of this book: how technology-in-use works to reconstitute our mode of being in the world, both directly and indirectly. Through an examination of how twentieth-century theorists and cultural movements have understood technology, the book attempts to articulate a position through which to engage with technology in the contemporary era.

This book explores the possibilities for a critical theory of technology. It does so by examining the work of various modern and postmodern thinkers and movements for whom the question of technology has been central. In the first part, the modernists Martin Heidegger, Walter Benjamin and the Futurist movement are considered. The second part examines Jean-François Lyotard, Paul Virilio and more recent forms of 'Cyberculturalism' as analysed through psychoanalytic theory. The book thus proceeds by way of a comparison of earlier modes with contemporary postmodern writers. Through examining the work of these theorists and movements, it aims to show, in both a descriptive and more distinctively theoretical sense, how technology impinges upon and reconstructs social and cultural meaning. I will argue that analysing this process enables us to reflect upon the conditions where on the one hand this transformation might be welcomed. On the other hand, however, it also allows us to make an argument about the need to set limits to the nature of technological mediation, an argument unavailable to theories based around a concept of neutrality, or to those that understand technology empirically. It is the aim of this book to engage theoretically with writings on technology in a manner which will enable us to go beyond the mere recognition that 'something' changes whenever it is technologically mediated.

Our relations with technology are often marked by a deep ambivalence. Intuitively, many people feel uneasy about the rate of technological change, yet can find no base from which to transform their feelings into a broader social critique. Such ambivalence can also be seen in the thinkers and movements examined in this book. Heidegger painted an overwhelmingly negative portrait of the baleful effects of modern technology. Yet even he always maintained that he was not simply against technology, that it was possible to say both 'yes' and 'no' to technology, and that technology contained its own 'saving power'. Benjamin both celebrated the technological destruction of 'aura' and lamented the loss of historical modes of experience that occurred through the technological reconstitution of human experience. Lyotard warns against the capacity of technology to lead to greater 'terror' and has noted the way that it colonises our notion of time, yet he looks to the techno-sciences and the information revolution to grant us access to a postmodern form of heterogeneous and libratory social relations. Virilio writes with both fascination and horror of the impact of technological change. Culturally, the Futurists hoped for a future based upon technological transcendence, yet their hopes were co-opted within a fascist aesthetic. Psychoanalytic approaches to cyberculture tend to either celebrate the subjective freedoms made possible through virtual environments or claim that such environments close off the possibility of meaningful subjective action. It is one of the aims of this book to examine the contradictions that emerge in the work of these theorists and movements, and to outline a social practice based in cultural reflexivity, which might move some way towards resolving, or at the very least responding productively, to some of these contradictions and tensions.

Given the Janus-faced nature of technology observed by all these writers, it becomes necessary, in formulating a critical theory of technology, to move beyond any sort of technological determinism. Determinism considers technology as a purely autonomous force that mysteriously shapes our way of being outside any socio-cultural context. We can see such determinism in narratives both of cultural pessimism and utopianism. Writers such as Ernst Jünger, Jacques Ellul, Baudrillard, and at times Heidegger, Virilio and Žižek, portray a world totally enframed by technology that precludes any alternative mode of engagement outside of its dictates. Such determinism also fuels the utopian sensibilities of the Futurists and many cyberculturalists, who argue that technology propels us towards a new evolutionary stage. Such sensibilities accept the existence of a unitary logic behind technology that drives us, for good or ill, towards an irrevocable future. This book rejects such determinism, arguing, alongside the best aspects of Heidegger, that we can say a 'yes' and a 'no' to technology.

To reject technological determinism, however, does not mean that we can simply approach technology instrumentally. The instrumental understanding of technology is based on the idea that it operates as a mere tool according to the subjective wishes of its users. Now while this common-sense notion may contain some truth, its truth must be radically circumscribed. This theory ignores the transformative role technology plays in reshaping and reconstituting subjectivity, embodiment and the social realm. To attempt a critical approach to technology from this position is all too often self-defeating, because it assumes that choices can be made from social and subjective positions which may themselves have been subject to a reconstitutive process. To say this is to acknowledge the manner in which technological mediation can frame social meaning in a necessarily different register. An example might be the way the Internet reconstitutes the social meaning of communication. As we all know, the Internet increases the sheer range of communicative possibilities by providing access to a wide scope of users. An instrumental approach stops at this point, regarding technology as simply extending the human capacity to communicate. This approach will only intervene in the case of a directly perceived possibility of harm, such as the availability of pornographic or other violent material. Yet one can go further, to note how the Internet constructs communication in a very specific manner. We can distinguish the sheer fact that communication is possible across greater extensions of space from the various qualitatively different modalities through which the meaning of that communication unfolds. While an instrumental approach will only consider the Internet empirically, I hope to consider such technology as part of a broader phenomenon that shapes the social in a new way. What does it mean if the social is increasingly constituted within a technologised setting, where the tangible presence of the other is no longer a structural necessity? In this context, we can consider the increasing phenomenon of Internet addiction, where subjects using the Internet are 'addicted' to sociality. What might it mean to be addicted to the social in this sense? An instrumental approach cannot ask these deeper questions about why something like Internet addiction occurs, because it fails to distinguish the qualitative differences between different modalities of the social.

One can think of numerous other examples in relation to emergent technologies: IVF reproduction, the possibility of human cloning, organ transplants. In each case, the instrumental approach only considers the technology in relation to the perceived harm or benefit to the individual. Broader social and ontological questions, such as the meaning of motherhood, the construction of human identity, the meanings of our own bodies, are ignored. These technologies are regarded as simply furthering already existing human capacities, the ability to create life, for example. The question of whether the meanings of these human capacities are *reconstituted* through the operation of a technological framework is not considered. This book will argue that this question is in fact crucial, and that understanding technology's capacity to reconstitute human meanings and activities within different constitutive frameworks provides the condition for determining whether we might say 'yes' or 'no' to technology.

How, then, are we to understand technology in a manner that will allow us to reflect usefully upon its relation to broader social and ontological questions and thus to move beyond an instrumental or empirical approach? Drawing upon the diverse points of view considered in this book, we can commence from the point that technology acts as a means of reconstituting the settings through which individual, social, and cultural meanings unfold. In other words, rather than simply furthering a particular capacity for action or representation, technology works to alter the grounds on which meaning attaches to any action or representation. Certainly, all the work considered here focuses upon how technology is able to alter historically embedded social and cultural meanings: Heidegger's *Gestell*, Benjamin's *aura*, Lyotard's *postmodern condition*, and Virilio's *speed* all concern themselves in some essential way with the reconstitutive capacity of technology. Similarly, the utopian fantasy that underlies the rhetorics of both Futurism and aspects of contemporary cyberculturism is driven by this very prospect of a technological reconstitution of our being.

However, it is possible to go further and to examine a common element that relates to this process of reconstitution. As a general proposition, I want to argue that *technology enables a more constitutively abstract mode of engagement with the world*.[1] What follows is a brief outline of how I employ the concept of abstraction, and how this provides the theoretical framework that will enable us to determine the grounds for a critical theory of technology.

The concept of abstraction relates both to the realm of ideas *and* to material processes. Intellectual practice functions, for example, through being able to stand outside a particular social and cultural setting. Intellectuals in their capacity as intellectuals interpret the world from a more abstract vantage point than people who live predominantly at the level of practical consciousness.

That is, intellectuals work through abstract ideas, and, just as importantly, the mode of their work depends on being abstracted from what they are trying to take apart. To speak of abstraction in this way *does not* entail an essential dichotomy between the abstract and the concrete: even if all thinking is in some basic manner abstract, we can still distinguish, following Sharp, between levels or degrees of abstraction. For instance, in the realm of ideas we can say that Newtonian scientific thought is more abstract than its Aristotelian predecessor, and that quantum theory is more abstract again in relation to its object (Sharp 1985: 54). Abstraction as a *material* process is perhaps more difficult to comprehend.

Before elaborating the theoretical framework through which I employ this term, some initial examples may help to elucidate the basic idea of abstraction as a materially lived relation. Marx provides the example of commodity exchange, where relations are carried out in a more abstract setting than in historically prior forms such as barter or reciprocal exchange. Commodity exchange functions so that the specific identity of the participants is no longer a structural necessity, unlike in earlier forms of exchange. In other words, commodity exchange abstracts from the particularities of the participant agents and involves a material abstraction of value. Similarly, the technological practice of writing constitutes a more abstract form of intellectual exchange than an exchange framed within the modalities of a face-to-face relationship. With writing, an individual's ideas and impressions are able to be conveyed despite their physical absence. This more abstract relation allows for certain possibilities to emerge, for instance that one's ideas can be taken out of context, quoted without permission and more easily misunderstood than in a situation not framed by co-presence. The case of writing on the Internet allows a further elaboration of this process, since one is free here to construct multiple personae that bear no concrete relation to a physical or historical identity. As a final example we can think of the technological process of in-vitro fertilisation, which enables the process of childbirth to occur in a more abstract setting, without the necessary presence of bodies, emotions, or even the identity of the participants.[2]

However, it is important not to proceed too hastily. I want to elaborate upon how Sharp's conception of 'constitutive abstraction' will be employed in this book. If I have already spoken of levels or degrees of abstraction, the meaning of these terms has not yet been sufficiently elaborated. The constitutive abstraction argument is generally explored in relation to two broad theoretical categories: *modes of social integration* and *ontological categories of existence*. These categories reveal, first, how technology comes to mediate and extend the means through which social relations are carried out and, second, how technology enables the reconstitution of the various categories of social being.

Modes of social integration

Social integration can occur at various degrees of abstraction, though it is important to emphasise from the outset that no social formation is constituted entirely at a single level. Following the work of Paul James, I have analytically distinguished three levels of social integration: namely the *face-to-face*, the *agency-extended* and the *disembodied* (James 1996: 19). These do not exist as pure forms: however it is important for the argument presented within this book to understand the means by which some levels work to contradict, reconstitute, extend, or erode practices that occur at other levels.

To speak of the *face-to-face* is to outline a mode of social integration which emphasises the importance of mutual and embodied co-presence. Historically, this mode is dominant in tribal and peasant society, where social existence is largely defined by tangible relations of co-presence. The face-to-face thus indicates a level of integration where embodied co-presence forms the dominant structuring principle around which social life is gathered. It is not defined by the nature of interaction. Under conditions where face-to-face relations predominate even actions or interactions carried out at a distance are governed by the sense of the presence of the Other. James writes that:

> [i]n this sense the modalities of co-presence bind absence. For example kinship based upon the existential significance of being born of a particular body into extended lines of blood relation is a key social form of face-to-face-integration. In social formations where kinship is fundamental to social integration a person is always bound by blood or affinity even after the dramatic separation brought by death.
>
> (James 1996: 24)

The face-to-face is the oldest and most basic form of social integration. More abstract modes of integration draw upon the meanings and experiences derived from this prior level. This is not to posit the face-to-face level of social integration as an idealised essence, but rather to recognise its historical and cultural embeddedness within social life. As Sharp puts it:

> a recognition of mutual presence as an irreducible ontological and ethical reference point is indispensable. The actual values concerned with this 'essence' are primarily a social construction: which, while relative to the degree to which it could have been engendered, is nevertheless the only essence our history and culture offers us. As a reference point it therefore carries with it a powerful imperative which cannot be totally renegotiated and certainly cannot be ignored
>
> (Sharp 1996: 6)

Similarly, when Heidegger asserts that Being-with-Others is a primordial relation (this is discussed in Chapter 2), and when Benjamin describes aura as the anticipation of another's gaze (Chapter 3), they are, I suggest, revealing the binding qualities of this mode of integration. In relation to technology, we can say that under conditions of the dominance of face-to-face integration, techniques and technologies enable people to engage with the world as an elaboration of the hand and the eye; and that they do so without fundamentally changing the nature of the world.

More abstract modes of integration draw upon, mediate and extend relations conducted at the face-to-face level. For instance, *agency-extended* relations, such as those made possible through institutions such as the church or state, or though practices such as commodity exchange, alter the context through which we engage with the Other. The Other becomes more *universalised*: a multiplicity of persons or things can function as a 'corporate' body in a particular structural location. However, despite this process of universalisation, under conditions of the dominance of agency-extended integration, techniques and technologies are used to extend the possibilities of human interrelation while still being bound by the limitations of embodied exchange.

At a further level of abstraction lies another distinguishable mode of integration: the *disembodied*. Technology facilitates the operation of this particular modality, in so far as it allows the overcoming of the constraints of embodiment, enabling the subject to engage with the world outside of a specific embodied location in time and space. At this level, the relation between self and other undergoes a qualitative shift. The tangible presence of another is no longer a defining dimension of the causal setting for interaction. It is at this level that we can see how technology dramatically reconstitutes social relations. The rise of the information society establishes a framework whereby communication is possible over greater extensions of space. Within this context, culture tends to be based around the image, and cultural objects can be consumed outside of the socio-historical locus of their production. Communities of strangers, most notably the nation-state, become possible, and relations formed in the dominance of the face-to-face tend to retreat into the private realm. In other words, the level of social integration is technologically mediated.

This mode of integration involves a contradictory process, whereby 'connection' takes place in a context of actual physical disconnection. Lyotard's notion of postmodern selves as 'nodal' points located within the flows of the information society reflects the emergence of this level of social integration, as does Virilio's somewhat exaggerated description of the masses as 'a multitude of passers-by' (Virilio 1996: 3).[3]

Within the terms of the levels of social integration analysis, it is important to distinguish an instance of a given relation from its ontological *modality*. If we take the earlier example of forms of tribal association, we can understand how relations of co-presence serve to *frame* the relation. In such a case, it is possible to see how the relation of co-presence is dominant over the more abstract possibilities offered by the exchange of gifts or the use of signs in communication. By contrast, we can imagine an ontologically different instance of face-to-face interaction, such as a meeting of executives flown in from different parts of the country for a one-hour meeting. In this case, despite the fact that the participants are co-present, as agents for their particular employers they are framed by a more abstract relation.

Moreover, it is also important to recognise that *in practice* different levels of social integration intersect rather than exist in analytic isolation. We can think of the infamous 'rape in cyberspace' (Dibbell 1993) The emotional trauma of being 'virtually' raped despite the physical absence of the rapist is indicative of the continuing relevance of the face-to-face modality within the nevertheless dominant level of disembodied integration.

Ontological categories of experience

Time, space, embodiment, and subjectivity form particular categories of experience through which we can examine the reconstitutive process technology enables. In this book they provide recurring themes through which to approach the question of technology. While all the writers considered here are concerned with one or more of these categories, I hope to draw them into a more comprehensive framework, which will establish their influence on the taken for granted ways of being and acting in the world.

Categories of space and time

From Kant we know that space and time are most usefully treated not as objective categories, but rather as malleable forms which shape our understanding. There is no doubt that the lived sense of these categories is fundamentally altered through the process of technological mediation. As Virilio notes, '[b]asically time is lived – physiologically, sociologically and politically – to the extent that it is interrupted' (Virilio 1991: 82). By examining the different ways time is interrupted, and thus lived, it becomes possible to distinguish more or less abstract relationships to time. We can distinguish a more concrete relation to time, where temporal experience is measured by lived routines based around daily and seasonal activity. For example, tribal and peasant societies based their activities around seasonal times, so that an activity such as planting seeds would occur according to phases of the moon. In saying this I do not intend to pose a simple dichotomy between traditional and modern forms of temporal experience, or to imply that tribal societies were framed entirely within a single temporal paradigm. A multitude of times co-exists within any social formation. I *am* suggesting that particular modalities of temporal experience can become more or less dominant within any given society, and that technology plays a crucial role in this process.

The introduction of technologies to measure time allows for a more abstract relation between subjective activity and temporal experience, as our activities are spread out over measurable and regulated units of time. At this level we work, play, eat and sleep in relation to discrete units of time. At a more abstract level still are the contemporary experiences that are less dependent upon time as an orientating principle. For instance, communications technologies allow for 'instant' access to people or things *at any time*. Stephen Kern notes how the rise of the first communications technologies, newspapers, telephones and telegraphs, allowed a greater sense of simultaneity, where a sense of the global first became apparent. The sense of the present was *thickened* when large numbers of people felt they were experiencing the same events (Kern 1983: 171). For Virilio, this is the level of time that threatens to uproot humanity from any meaningful sense of worldly engagement.

There are other examples that indicate this more abstract relation to time. For instance, the changing nature of work practices increasingly offers 'flexitime', so that an employee can work the requisite number of hours at any time of day or night. The rise of the consumer society has led to the increase of 'services' such as twenty-four hour, seven-days-a-week shopping or banking. These possibilities are extended as consumer transactions become increasingly computerised.

Such transformations work to detach human activities from any naturally grounded notion of time. Heidegger's metaphors of 'standing reserve' and the 'age of the world picture' help to evoke this scenario. When something is in standing-reserve it is in a sense 'outside' time, ready to be harnessed at any time. Envisioning 'the world as picture' also conveys the sense of an increasingly image-based society where cultural signifiers avail themselves outside of the restrictions of a particular time or place. This more abstract relationship to time assumes an increasing degree of dominance in capitalist societies which entrench a means-ends paradigm. The result of an activity is seen as outside of the activity itself. Consequently, the time taken to achieve the end or goal can be technologically accelerated. Time is no longer constitutive of the act. This is a concern Lyotard raises in his work on time.

Similarly, we can think of more and less abstract relations to space. At one end of the spectrum we can think of activities that can only take place in relation to specific spaces: the tribal understanding of the 'sacred site' provides an example. New technologies, on the other hand, make the relation between an action and the space in which it unfolds more abstract. Benjamin's notion of how technologies of reproduction facilitate the masses' desire to 'bring things close', as the uniqueness of reality is overturned through the acceptance of its reproduction (Chapter 3) serves as a case in point.

The introduction of electronic communications technologies in the last twenty years has allowed subjective actions to become increasingly detached from any specific spatio-temporal location. William Bogard notes how:

> Mainframes and PC's, automated tellers, barcodes, vision systems, faxes, xeroxes, telephones, microwave relays, electromagnetic sensors, video games, expert systems processing information at the speed of light, transcending the constraints of time and distance, these devices increasingly supplement and substitute for the slow, localized process-control technologies of the industrial period.
>
> (Bogard 1996: 15)

This more abstract relation enables us to engage in specific activities no longer anchored to any particular time or place. As we shall see, Howard Rheingold's desire for virtual spaces that might envelop community, in the same way as less abstract spaces like the village square or town hall, ignores the crucial issue of how a more abstract relation to space alters the meaning of perceptions and actions that unfold within it.

Subjectivity

This category examines the changing nature of the settings that constitute subjectivity. We have already noted the way in which identity in tribal forms of association is determined with regard to a binding range of socio-cultural determinants, such as tribal history and kinship rules. Within modernity, if these settings have expanded, the subject is largely constituted through a series of mobile-but-stabilising social and cultural frames. These settings, embedded within the practices of social life, function as a source of both identity and authority, constituting what has come to be known as the 'centred' subject of modernity (Poster 1990: 14). This level, which allows mobility within a larger context of stability, is what simultaneously conditions and contains the Futurists' ideology of transcendence, to the degree that their technological imagination reinvigorates prior fantasies within a context that, if established, would actually undermine the conditions for their realisation (Chapter 4).

Within postmodernity, the dominance of this mode of subject formation has declined. This transformation can be traced through the shift in the individual–society relation.[4] If the social settings of modernity carried with them a certain imperative force, which at once constrained the individual and at the same time granted symbolic recognition to their actions, the postmodern shift towards a network society, where the subject is constructed at points of intersection between systems of signification or technological flows, causes a decline in the imperative force of these earlier settings. Instead, the imperative is placed on the individual to become increasingly autonomous and to shape their own lives. To speak of 'autonomy' here is to refer to a specific social and historical form of the subject, rather than the 'morally autonomous' subject of classical liberalism. Autonomy refers to the historical shift whereby subjects must radically constitute themselves in the absence of social and cultural roots or determinants.

Gerry Gill traces this transformation:

> [t]here has been a shrinking of this centre of the life world from the community to the nuclear family, with a simultaneous expansion of that centre from the parochial boundaries of the community to the more universal national and international culture. To a large extent this mode of connection is . . . [an] abstract 'mediated' one enabled through mass media . . . the person of today must increasingly go to a more abstract level for cultural signs and meanings . . . [they] must

increasingly develop a more abstract sense of self: one that stands outside of particular roles, norms or signs and is responsible for making its own identity.

(Gill 1984: 91)

The move from a modern to postmodern mode of subject constitution creates a shift in the disciplinary structures that envelop the subject. As Gill points out, the postmodern subject, in the absence of the more overarching structures of modernity, must discipline themselves in order to obtain cultural material for their own identity. It is at this more abstract level of subject constitution that we can locate Heidegger's critique of the self-framing mode of technological subjectivity, and in a more radical sense Lyotard's 'nodal' subject. Cyber-technologies either provide radical possibilities for subjective action, a claim made by Sherry Turkle or, according to Slavoj Žižek, they close off the ontological gap which underwrites subjectivity.

Embodiment

Different forms of embodiment have inscribed themselves within history and across various social formations from tribalism to postmodernism. Less abstract forms of embodiment understand the body as a relatively stable signifier intertwined within a specific range of social practices. Thus, at this level the body can serve as a marker of race, age, or gender, the social and cultural meaning of which is relatively fixed.

More recently, we can trace the reconstitution of our lived sense of embodiment in contemporary society, where the social meanings of embodiment are more arbitrarily constituted. Indeed the body itself is no longer regarded as a fixed symbol of identity, but is subject to a variety of technological practices (such as plastic surgery or the gymnasium) which enable the reconstitution of the 'natural' body into a more open network of signification (James and Carkeek 1992). Lyotard's theoretical destruction of the organic body, replaced by a postmodern body composed of separable and reconfigurable libidinal zones, is indicative of this shift.

Other technologies work to reconstitute our lived sense of embodiment, in the sense that they attempt to overcome the material limitations of the 'natural' body. From the fortified warrior subject, fantasised by the Futurists, to the virtual traveller surfing the Internet, the technological transcendence of the body has become an increasingly widespread phenomenon.

Any discussion of these categories will remain ineffective unless it can then refer back to the social forms which carry them. The term social form is used here to indicate the broader context in which any action or belief derives its value, and more generally to the way social relations are structured. At any given historical period the broader social formation consists of these different forms, held together in an uneven and conflicting relation.[5] Having established that the social consists of an intersection of these forms, it is time to return to the more specific issue of how technological practices can be understood in relation to them.

If technology allows for a more abstract mode of engagement with the world, if it reconstitutes social and cultural settings, in effect creating a new constitutive framework through which to operate, then how ought we to negotiate the relationship between this emergent level and prior levels of engagement and association?[6] The question this book attempts to engage is: what are the effects of this modality of practical abstraction becoming dominant as a way of relating to others? Technology is often constructed as a 'tool', a means of resolving human needs. Yet what happens when those needs and their attempted resolution are played out within a more abstract framework constructed through the process of technological mediation? What happens to social meanings when they intersect with technological practices? If the framework through which meanings are attached to subjects, actions and cultural contexts is reconstituted through technology, is it possible (or desirable) to envisage an alternative mode of practice that can engage with technology without going over fully to a mode of being where social life will be entirely constituted within the terms of this more abstract framework? Such a technologised and abstract framework would mean, to return to the example of the Internet, something like spending our whole lives framed within the terms of the abstract community of strangers, even when we were not using the net. To be framed in such a way would mean that our social relations would take on a fleeting and transient quality, even within a situation involving actual co-presence. The binding qualities of co-presence as a modality of social engagement would thus have been reconstituted within the terms of the Internet. Is it possible to constitute social reciprocity and human co-operation within such a framework? I will attempt to answer this question in the body of the book.

The argument

Each of the following chapters deals with theorists or cultural movements that have concerned themselves with the question of technology. All the writers considered have to varying degrees recognised the ambivalent promise of technology, often combining a critique of technology with a particular strategy for a different relationship with it that might open up radical possibilities for the reconstruction of social life. For analytic purposes, I have divided the book into two broad sections, modernity and postmodernity. This division is grounded historically: Heidegger's work, or at least the aspects I will focus on (Chapter 2), Benjamin's work (Chapter 3), and the Italian Futurist movement (Chapter 4), all occurred in the first part of the twentieth century. With the Futurists we can see the first cultural movement to base itself around a direct affinity with technology. They embraced the appearance of new technologies precisely for their ability to enhance new modes of being in the world, for instance the aeroplane and the motor car and their relation to the cult of speed. For the Futurists, technology functioned as a means of *transcendence* of the structures of everyday Italian culture and society. This transcendence is achieved precisely through the technological reconstitution of life at a more abstract level. Heidegger, too, understood technology as a reconstitutive process, though he drew a far more pessimistic conclusion than the Futurists. Indeed, Heidegger is far more profound, in that he questions the very ground from which any technological transcendence could occur. His analysis of technology is, I suggest, the most comprehensive of all writers on the subject. In effect, he sets up a position from which to judge the other theorists and movements. To what extent do they manage to engage with technology in a manner that allows them to escape the terms of the Heideggerian problematic? While Benjamin provides no 'theory of technology' in the comprehensive sense of Heidegger, he nonetheless attempts to theorise a far more productive relationship to it. And, unlike the Futurists, he provides the first steps towards saying 'yes' to technology outside the terms of the *Gestell*. In order to reveal how this is possible, I read Benjamin's work through the lens of the constitutive abstraction argument discussed above and through a discussion of the social forms that carry varieties of auratic experience.

I have already related how social formations involve a number of intersecting and conflicting modalities of being and acting, each constituted at a different degree of abstraction. If technology enabled a more abstract mode of worldly engagement within classical modernity, this particular modality was still only emerging and had yet to assume the more central position it has in our own time. Take, for example, the way the cinema was understood to hold a radical potential for both Benjamin and the Futurists. For both, it constituted a specific *site* for certain experiences, such as narrative displacement or mimetic shock. In late modernity or postmodernity, by contrast, rather than cinema forming a specific bounded site where a certain type of experience is possible, life in general has become more 'cinematic', as experience is increasingly constituted through the consumption of images, bypassing the modalities of co-presence, and becoming more fragmented and technologically mediated. The metaphor of 'life as cinema' is especially relevant given the growth of technologies which harness 'the mobile and virtual gaze',[7] such as virtual reality, computer games, the shopping mall, the experience of driving on a freeway and so on. Such a metaphor expresses a tacit recognition of this reconstitutive process whereby this more abstract level of engagement has assumed a relation of dominance with regard to other, less abstract ones. To talk of postmodernity then, is not to invoke an epochal shift, whereby all that is modern is simply swept away. Rather, it is to examine the increasing centrality of a technologically reconstituted and abstract layer of social being.

The situation we find in postmodernity is that the technological mediation of the received categories of social life is so widespread as to allow the more abstract ways of being and acting in the world to become increasingly dominant. In other words, the quantitative change in technological mediation also entails a *qualitative* change. If, within modernity, the broadcast news of the world made an impact on the local community, then the social formation remained largely constituted within a less abstract modality than that governed by extended media relations. Within postmodernity, however, the increase in media and communications technologies means that social integration is increasingly carried out at the level of disembodied-extended integration. The dominance of this level reconstitutes the way we engage in less concrete modalities. Take the distinction between the local and the global. Within the dominance of a more abstract mode of engagement, the local begins to operate in the same abstract framework as the global. Anthony Giddens describes this process as 'reality inversion', whereby:

> [t]he transformations of place, and the intrusion of distance into local activities, combined with the centrality of mediated experience, radically change what 'the world' actually is. This is so both on the level of the 'phenomenal world' of the individual and the general universe of social activity within which collective social life is enacted. Although everyone lives a local life, phenomenal worlds are for the most part truly global.
> (Giddens 1991: 187)

Giddens's observation marks this reconstitutive process whereby the qualitative differences between the local and the global decrease. For example, the tourist industry encourages us to explore the local in the sense that it promotes cultural uniqueness, eccentricity and difference. However, within any such context the local is increasingly grasped through marketing images, signs and information: at this level it merely forms part of a globalised databank.

The relative dominance of this more abstract mode of being brings with it novel theoretical approaches to technology. For instance, Lyotard theorises a mode of social relations around the temporary contract, computer networks, and legitimation through paralogy (Chapter 5), while Paul Virilio argues that time and space are now threatened 'habitats', a claim which resonates in a way quite impossible fifty years ago (Chapter 6).

In analysing the thinkers and movements discussed in this book, I endeavour to register the importance of certain aspects of their work in relation to the way they attempt to come to grips with the dilemma of technological reconstitution. Yet ultimately each theoretical account remains limited. By theorising the question of technology through an account of its relationship to constitutive forms, I hope to outline a more productive strategy. Such a strategy enables us to move beyond Heidegger's pessimism without recourse to the ethereal strategy of *Gelassenheit*.[8] It allows us to take up the scattered speculations upon technology in Benjamin's work, to develop them within the theory being outlined here, and thus to move beyond more naive readings of his work as simply celebrating the technological destruction of aura. It allows us to understand the Futurist expression of aggressive ideologies through technology. It enables us to go beyond Lyotard's false dichotomy of totalisation and heterogeneity and the one-dimensional critique of Virilio. Finally, I explore how psychoanalysis has theorised virtual technologies, especially how the subject is reconstituted within cyberspace. I attempt to steer a course between the pessimistic analysis made by Žižek and the more utopian sentiments expressed by Turkle and others. Instead of alternating between totalising ontological approaches and more pragmatic political projects based around radical performativity, I argue that a differentiated ontological approach is possible, one that would enable a politics of (virtual) becoming, but would remain sensitive to the wider contexts which would govern any production of subjectivity. I argue that Teresa Brennan's work suggests how such an approach might be taken (Chapter 7).

This book will argue that there is a need to understand the dialectical interplay between different forms of social life, some of which are grounded in stability and concrete interactions with the world, while others arise within a more abstracted setting that technology enables. This follows from the work of Sharp in envisioning the social as composed of an intersection of constitutive layers (Sharp 1985). Such an approach recognises the importance of relations structured in mutual presence and tangibility as the basic carriers of social meaning yet remains open to the radical possibilities contained within a more open and abstract constitutive form.

To the extent that they attempt to analyse the construction of meanings within this more abstract level, this book recognises the contribution and importance of poststructuralist and postmodernist modes of thinking.[9] However, it also recognises the limitations of these models, in that they overgeneralise the degree to which meanings are constructed within this level. The contributions of Lyotard, Baudrillard, Wark and the sensibilities of many proponents of cyberculture are important for analysing radically new ways of being and knowing within this emergent framework. Nevertheless, they tend to overgeneralise the extent of the transformation, treating the more abstract forms of social integration and experience as constituting the whole of the social formation. The problem of overgeneralisation is, in this context, highly significant and not confined to the mere exaggeration of social effects. On the contrary, it cuts off access to a strategy that might be able to retain the benefits of technology without the need to remain entirely within the more abstract modality that frames these conceptions of the postmodern. As we shall see in Chapters 5, 6 and 7, which cover Lyotard, Wark and cyberculturalism respectively, such theories tend to posit a choice between totalitarian or heterogeneous possibilities, where technology provides radical freedoms only to the extent that it is able to collapse the entire framework of social meaning into an open circuit of fragmentation and difference. If the radical emancipatory politics of many poststructuralist and postmodern theorists are limited by the extent to which they still adhere to a conception of life as occurring within a single plane of being,[10] so too is the more overtly critical politics of Paul Virilio, who remains unable to formulate any kind of strategy except denunciation against what he regards as an overwhelming technological threat to our existence.

I want to argue that we do not have to remain locked into the dichotomy of heterogeneity and totalitarianism. It is not that these analyses are inaccurate, but rather that they either understand social life and politics as possible only within the abstract social form facilitated through technology, or, despairing at such a politics, hold to a heroic individual denunciation of what they regard as an oppressive and overwhelming technological imperative. It is the contention of this book that there is a third path. This would require a certain cultural *reflexivity*, recognising the benefits of technological reconstitution while setting limits to the extent of its operation.

How might such limits be formulated? If we understand technology as enabling the constitution of a more abstract mode of engagement with the world, and as operating through the reconstitution of the settings from less abstract levels, then we can begin to develop criteria through which to decide whether to welcome or reject this operation. As a general proposition, it is possible to identify two principles that provide strong justification for setting limits to the framing of life entirely at a more abstract 'technological' level. The first concerns the degree to which the abstract reconstitution of social and cultural meanings is easily harnessed to the commodity relation. For instance, if communications technologies offer us the possibility of keeping in 'touch' at a distance, they do so at the cost of selling social integration back to the consumer through time charging, or the continual need to purchase software in a marketplace based on rapid and continual obsolescence. If personal identity-formation is able, through technology, to reach out beyond the local to the global in the form of information and image, it must be remembered that this extended process of identity-formation is all too often constructed through the purchase of commodities and the meanings attached to them. The second criterion is related, but concerns itself with the broader questions of social and cultural meaning I discuss under the rubric of *ontological contradiction*.[11]

We can define ontological contradiction as the process whereby desires and practices contained within one constitutive layer contradict those carried within another. Insofar as human needs and desires are carried within specific historical and cultural frameworks, it is necessary to consider whether technology is able to consummate these needs, or whether the reconstituting process it enables works to undermine the ground which historically sustained them. Several examples of ontological contradiction can be briefly invoked from within the body of this book. Heidegger's technologised subject, whose power to objectify the world through a process of abstraction only comes at the cost of objectifying the self, is a classic formulation. Paul Virilio's claim that technological 'speed' ultimately leads to inertia has a similar tenor, in that both theorists describe a situation in which technological mediation extends previous capacities only to undermine the ground on which such extension could have any meaning. The Futurist desire for a more powerful and regenerated nation-state was attempted within a theory that worshipped technology for its transgressive and universalising character. Yet the nation-state was invoked at the very same time that emerging technologies allowed for the easy transcendence of any physical and cultural boundary and hence threatened to undermine the meaning of the nation. As a final example we can think of the contradictions that might ensue when attempts are made to transcend material oppression (such as racism or sexism) by adopting a different persona on the Internet. Does this actually solve the problem which has occurred at a prior level of experience? To what extent can it do so, if it comes only by shedding one's more tangible identity because of the problems associated with it?

It is not the intention of this book to resolve these contradictions in some sort of grand Hegelian manoeuvre. Indeed, if we were to affirm reflexively the importance of more concrete frames of reference over postmodern forms of openness and heterogeneity, then this in itself would result in a contradiction, in that the openness and radical choice enabled by postmodern forms of life would, to some extent, be constrained. Any ethical question inevitably involves an aspect of choosing. The important point is not to resolve the contradiction, but to make a choice that can best lead to the re-establishment of a co-operative ethic within contemporary life.

Understanding technological reconstitution is of vital importance for a reflexive approach, yet is rarely considered so in critical theories of technology. Thus, if we can agree with Andrew Feenberg that '[i]f an alternative modernity is possible, it must be based not on contents but on deeper differences in cultural forms' (Feenberg 1995: 214), then we must recognise the need to explore further the relation of any such forms with technology. When Feenberg goes on to argue for the social framing of technology, insisting that '[t]echnologies are always already invested with aesthetic, ethical and cultural values' (Feenberg 1995: 229), he elides the question of how the technological reconstitution of such values can give rise to ontological contradiction. If, for instance, an argument for the widespread use of IVF stems from the wish to gratify the human desire for childbirth, then to what extent does the industrialisation of the birth process erode the broader, historically embedded settings which generate the original desire? (White 1981: 29) It is not my intention to answer specific questions such as this, but rather to argue for a framework through which they might most effectively be asked.

2 Beyond enframing
Heidegger and the question concerning technology

> The works that are being peddled about nowadays as the philosophy of National Socialism ... have nothing whatever to do with *the inner truth and greatness* of this movement (namely the encounter between global technology and modern man).
> (Heidegger *Introduction to Metaphysics* 1935)

> Man as a rational being of the age of Enlightenment is no less a subject than is man who grasps himself as a nation, wills himself as a people, fosters himself as a race, and, finally empowers himself as lord of the earth.
> (Heidegger *The Age of the World Picture* 1938)

> 'How could you think that a man as uncultivated as Hitler can govern Germany?'
> 'It's not just a question of culture. Take a look at his wonderful hands!'
> (Exchange between Karl Jaspers and Martin Heidegger 1933)

To date, Heidegger remains one of the most influential philosophers of technology. His understanding of how technology reconstitutes our relationship to the world in many ways informs the theory of constitutive abstraction I am working with in this book. Heidegger's importance lies in recognising the role of technological mediation in shaping our modes of engagement with the world. For Heidegger, technology cannot simply be conceived as merely this or that machine or device, but must be examined through the way in which it works to enact an ontological shift. This recognition is essential as a means to developing a critical relation to technology. The alternative, to conceive of technology as merely neutral and therefore subservient to other value spheres, is to ignore the role that technology plays in reshaping our relations across the social realm. Any theory which seeks to engage with technology on the assumption that it is a mere tool to be manipulated for substantive ends, remains naive. As I have already argued in the introduction, such a position presumes to treat technology from a prior framework of understanding, thereby ignoring the reconstitutive capacity of technology. In sharp contrast to this position, Heidegger recognised that our understanding of the world is radically altered in the age of technology. From Heidegger we can derive the important point that we need to comprehend thoroughly the new ontological framework that is enabled through technology before we can even begin to renegotiate our relationship with the world. Hence, simple calls for a responsible use of technology, or theories of technology which welcome its use when it 'extends' human capacities, remain limited because they ignore the reconstitutive capacity of technology to alter the conditions which grant significance to any action or value. Because he theorised technology precisely through its relation to questions of ontological framing, Heidegger provides a valuable starting point.

Despite its enormous significance, Heidegger's work remains problematic. These problems can be grouped around three related areas. First, his theory of technology is overdetermined, leaving no room for an effective response to technological domination. The second problem concerns his entanglement with National Socialism and the extent to which his work can overcome the fascism that, on one level, he heartily endorsed. The third concerns the extent to which his preoccupation with Being reveals what some critics have called a lack of ethical responsibility, which necessarily limits the force and capacity of his insight.

In relation to the most commonly perceived problem, that Heidegger leaves no space for effective resistance to technology, I want to argue that this is only partly true. By way of illustration, I outline the most common critiques of Heidegger's later work and, while concurring with them up to a certain point, I claim that Heidegger's understanding of technology as an epochal shaping force is more profound than his critics allow, and certainly ought not merely to be dismissed in the light of his complicity with National Socialism. Nevertheless, I examine the problem of a levelling process in Heidegger's work, which fails to distinguish adequately between different kinds of phenomena. In relation to this, I analyse Heidegger's notorious statement concerning the essential equivalence between the Holocaust and the rise of mechanised agriculture, in order to question his prioritisation of questions of Being over questions concerning the fate of particular beings.

With this in mind, I examine the strengths and limits of approaches that attempt to supplement Heidegger's ethical 'indifference' by formulating a more comprehensive and ethical account of the concrete Other, such as the attempt to account for the embodied nature of human being in the world. I argue that while such approaches are insightful, an ethical approach which accommodates the thinking of the concrete Other can best be approached through a return to the question of constitutive framing, albeit in a more reflexive manner than Heidegger was able to achieve. The final section of this chapter argues for a plurality of such frames and examines neglected aspects of Heidegger's work, especially his conception of other constitutive forms such as politics and Being-with-Others. I conclude by suggesting how Heidegger's work could be drawn into the framework of the constitutive abstraction argument. Heidegger's analysis of different modes of revealing can contribute to a layered social ontology; which as Chapter 1 argued, is essential for allowing a more reflexive approach to technology. Such an approach would go beyond Heidegger's pessimism and ethical indifference.

Heidegger's theorisation of technology through its essence

Heidegger claimed that the way we understand ourselves, and how other entities reveal themselves to us, were radically altered by technology. In his essay 'The Question Concerning Technology', he considers how we might establish a 'free' relationship to technology. By this he means a way of remaining open to the possibilities of technology, whilst escaping the metaphysics of subjectivism and domination that technology tends to facilitate. Immediately, then, Heidegger's position can be distinguished from that of the Italian Futurists, who regarded technology as simply generating a vast spectrum of possibilities which enhanced freedom. In other words, the Futurist's equated technological *possibility* with freedom. When Heidegger speaks of freedom, it is of a freedom that is radically circumscribed by the ontological framework through which freedom is made possible. Unlike the Futurists, Heidegger's sense of freedom is never simply equated with individual empowerment or transcending limits. How, then, would Heidegger have us respond to technology?

He begins by examining the 'essence' of technology. This essence is not technological, but rather, concerns a particular mode of revealing. This is a crucial move that goes beyond the consideration of any particular technology so as to focus on how technology both shapes and itself is situated within a broader constitutive framework. For Heidegger, then, technology functions as a frame of reference which constitutes the manner in which things come to presence. The way things present themselves before us constitutes the manner of our worldly engagement. On this point Michael Zimmerman notes 'if things manifest themselves as creatures of God, people treat them one way, if things reveal themselves as nothing but raw material, people treat them in another way' (Zimmerman 1990: xv).

Heidegger finds that the technological mode of revealing takes the form of an aggressive 'challenging', a provoking which causes us to challenge everything, including ourselves as an objectified entity, ready for technical manipulation. Technology constitutes a framework which reduces our phenomenal mode of engagement with the world to understanding it through the narrow parameters of use-value. This readiness-for-use Heidegger calls 'standing-reserve'. Constituted as standing-reserve, things become objectified, caught up in the giant 'ordering process' that characterises the modern world: they are stockpiled, ready for use in technological projects. It is this mode of revealing of the entire world as standing-reserve that allows human beings to dominate *through* technology. In other words, it is at first necessary for us to view the world as standing-reserve before we can dominate it through the use of modern technology. For Heidegger, this technological mode of revealing changes our relationship with the world. He claims that:

> even the Rhine appears to be something at our command ... the river is dammed up into the power plant. What the river is now, namely a water power supplier, derives from the essence of the power station.
> (Heidegger 1993: 297)

Heidegger claims that technology radically alters the meaning of phenomena. Things are able to be removed from the contexts through which we once understood them.[1] The cause of this transformation stems from the essence of technology. The mode of revealing which views and orders the world as standing-reserve is called the *Gestell*, the setting that enframes. Because the enframing is a constitutive setting, and not any particular technique it is:

> nothing technological, nothing on the order of a machine. It is the way in which the real reveals itself as standing reserve ... [e]nframing is the gathering together which challenges man and puts him in position to reveal the actual, in the mode of ordering, as standing-reserve.
> (Heidegger 1993: 329)

While Heidegger does not use the term very frequently, we can observe how the mode of revealing constituted through the *Gestell* is made possible through a radical process of abstraction. Technology enables things to be lifted from the more concrete frames of reference through which they had previously revealed themselves to us. This abstracting process is linked with a kind of 'possibility explosion',[2] a sensibility that permeated the imagination of modernity and, as we shall see, was central to the Futurist project of individual and cultural renewal. The removal of restraint made possible through new technologies has its most profound effect, for Heidegger, in relation to the categories of space and time.

Heidegger observed that technology worked to promote a shrinking of the world. While technology enabled a breaching of distance, it also concealed other forms of interaction which take place within the world. Heidegger argued that the essential nature of things lay in their 'nearness'. Nearness involves how things reveal themselves as meaningful within the context of our dealings with them. Instead of bringing things near, the shrinking of distance and time by technological means has caused nearness to withdraw. He explains this paradox thus:

> [a]ll distances in time and space are shrinking. Man now reaches overnight, by plane, places which formally took weeks and months to travel ... The peak of this abolition of every possibility of remoteness is reached by television, which will soon pervade and dominate the whole machinery of communication. Man puts the longest distances behind him

in the shortest time ... Short distance is not in itself nearness. Nor is great distance itself remoteness.

(Heidegger 1971: 165)

Heidegger distinguishes the facticity of distance from its ontological modality. The abolition of distance achieved through technology brings us no nearer to other beings or entities. They lose their nearness and become objectified as they are removed from their traditional context, of which meaningful distance formed an integral part. Technological modes of extension increase the tendency for us to see, and conquer, the world as picture. While technologies such as television and cinema tend to give Heidegger's metaphor a literal quality, it is meant to indicate a much broader and more profound phenomenon, referring to the process of taking hold of the world *outside* of any particular technology. Emerging ways of being in the world, such as transport technologies which allowed one to travel greater distances, or media forms which allowed the events of the world to unfold almost instantaneously in one's own home led into a perception of the world as something eminently graspable, understandable and therefore controllable. The transformation of the world into a pictorial representation constitutes, for Heidegger, 'the fundamental event of our age' (Heidegger 1977: 133). The notions of 'local' and 'distant' lose their groundings when the world becomes representable through the totalising vision of modern humanity. The shrinking of the world through technological mediation changes how the world reveals itself. It can now be objectified, subjected to calculation and interpreted as standing-reserve. This can only occur through a more abstracted engagement with the world. Technology works to eradicate all remoteness, yet the 'closeness' of once distant things is only possible through their abstraction as pictures. Heidegger links this process to the drive for domination that manifest itself through *Gestell*. As Michael Zimmerman observes '[t]he attempt to make everything close and available arises from the increasingly one-dimensional ontology of modernity: everything appears to be nothing but various kinds of matter which can be used and switched about at will' (1990: 151).

Thus, from a Heideggerian perspective, one can operate through a mode of being that is 'technological', regardless of whether one is actively involved in using any particular piece of technology. One could take the example of certain strands of the New Age movement, which appropriate the artefacts of distant and 'exotic' cultures in order to grant some aspect of 'spirituality' to their own practices. Despite often being resolutely anti-technological in the narrow sense, it is clear that their worldview is framed by the *essence* of technology, in that other cultures are brought close and available, perceived only through the context of use-value and reconstituted through a one-dimensional context that valorises the universal quality of the spiritual, regardless of the locality or initial context that embedded the meanings surrounding that culture's artefact. Distant cultures are made close, but they are reconstituted as standing-reserve in the process. To counter the effects of technological enframing would require that we go beyond an instrumental understanding of technology and, with Heidegger, examine the broader frame through which technology brings things into presence.

However, if Heidegger remains relevant for the contemporary era, he also shares some of the limitations of much of contemporary postmodernism. Heidegger's observations on space and time clearly form a precursor for the work of people such as Virilio and Baudrillard. When Virilio claims that 'when things "far" are brought into immediate proximity, those that are proportionately "near" such as friends, kin, neighbours – turn what is proximate – family, work, or neighbourhood – into a foreign, if not inimical space' (Virilio 1993: 10–11), he echoes Heidegger in acknowledging that the meanings of proximity and distance are not tied to their measurable quality, but rather reveal themselves within a particular constitutive framework. Yet we need to remain at a certain distance from the conclusions they draw from such insight. Both writers tend to give a one-sided reading of this reconstituting process, all too often seeing technology as leading to an inevitable nihilism. For instance, in a 1966 interview Heidegger remarked upon his own horror at seeing pictures of the earth from the moon on television:

> I do not know whether you were frightened, but I at any rate was frightened when I saw pictures coming from the moon to the earth. We don't need any atom bomb. The uprooting of man has already taken place. The only thing left is purely technological relationships. This is no longer the earth on which man lives.
>
> (Alter and Caputo 1976: 277)

The technological framing of the world as a picture to be grasped and manipulated by a totalising viewpoint, culminates in this image of the world manifesting itself literally as a television picture. For Heidegger, this leads to despair since the world is reduced to a representation that appears before the human subject. Yet, one surely needs to ask whether there is not some fundamental *ambivalence* surrounding this image? While it can be read as the culmination of the type of understanding which aggressively grasps the earth as a picture or more broadly as standing-reserve, it can also be read through a very different framework that would understand the image as signifying the uniqueness, finiteness and fragility of the world we inhabit. Even the horror Heidegger feels arises necessarily from a perspective outside of the *Gestell*. It is this unwillingness to explore the ambivalence of technology, and the ontological contradictions that result therefrom that prevents us developing a more effective response to technology. The question remains as to how we can develop the conditions that would make the latter form of understanding unfold more readily than Heidegger would deem possible. How to respond to this question is a fundamental concern of this book.

Despite this reservation, Heidegger's insight that our changing relationship to time and space alters our mode of engagement with the world remains of vital importance. The overcoming of embedded frameworks such as time and space allows for a different mode of engagement with the world, one that posits the subject as the source of all meaning. If the Futurists celebrated this emergent mode of subjectivity made possible through the technological overcoming of limits, Heidegger questioned the meaning and consequence of any such freedom. He argued that the technological overcoming of limits could only result in a crisis of meaning for the subject.

Subjectivity

Heidegger sought to understand the conditions that allowed technology to function as a means of domination. An essential condition for him was the technological reconstitution of subjectivity. He argued that technology worked to affirm and promote a particular mode of subjectivity that was aggressive and perceived itself as God-like. Heidegger understood the technological age as the culmination of a metaphysics begun almost two thousand years ago. This metaphysics grew to privilege the subject-centred understanding of the world. Through the assistance of technology, the modern subject came to experience the world as an ego-logical construct. It was through understanding how the modern subject functioned as the absolute point of reference that Heidegger linked technology to a modern form of nihilism. Because technology helped to remove certain 'natural' limitations, it created the illusion of limitless possibility, which, when combined with the fantasised omnipotence of modern subjectivity, worked to vanquish all traces of otherness. Heidegger writes that '[i]t seems as though man everywhere and always encounters only himself' (Heidegger 1977: 332). The impoverishment of the other also profoundly affected the ontological significance of any *possibility* made available to the self-framing subject.

Under the projections of technological subjectivity, all things become available for use. *Gestell* frames the meanings of the modern world, revealing it as stockpile of resources, waiting for humankind to shape it. Heidegger argues that this not only threatens the earth, but empties out the categories which underlie the initial desire for the transcendence of earthly limits. The increased scope for subjective power and manipulation made possible through technology carries with it a logic that erodes the resources from which the meaning of any subjective action could be granted. Ultimately, the subject flounders in an infinite number of meaningless possibilities, capitulating to the logic of domination by taking itself as standing-reserve in a constant effort to reshape itself. Heidegger describes the meaningless momentum of the modern world: '[e]verything is functioning. This is exactly what is so uncanny, that everything is functioning and that the functioning drives us more and more to even further functioning' (Heidegger 1981: 53).

Under *Gestell*, humanity can only provoke itself into further activity, subject to a spiralling logic of endless consumption and distraction. For Heidegger, the increase in subjective autonomy can only occur through a corresponding alienation from the grounding practices which constitute a meaningful relation to the world. Activity itself becomes technologically reconstituted, reduced to its degraded mode of 'functioning'. This is an important distinction that will be drawn upon again in relation to the discourse which surrounds contemporary information technologies. Instead of merely invoking the notion of activity, as many enthusiasts of cyberculture tend to do, we would do well to ask, in a Heideggerian vein, what constitutes meaningful activity? What distinguishes activity from mere functioning? Merely to rely on the possibility for subjective activity in relation to any technology invites the possibility of ontological contradiction.

Heidegger claims that subjective activity under *Gestell* merely 'provokes' the subject into a meaningless functioning. Consequently, the desire for subjective power, liberation or transcendence culminates instead in an erasure of the conditions through which such ends could be sustained. According to Heidegger, the objectification of the world, often a condition which underwrites the possibility of more naive forms of technological liberation, results in an objectification of the subject. For Heidegger, 'he comes to the point where he himself will have to be taken as standing-reserve' (Heidegger 1977: 332). Such a process results in what I referred to in the Introduction as an ontological contradiction. The conditions which make possible the technological 'fulfilment' of needs result instead in the emptying out of the ontological categories of meaning which granted significance to the initial desire. In the present example, individual power comes at the cost of self-objectification. Yet while Heidegger successfully draws attention to the fact that such a contradiction occurs, I want to argue that he does not adequately respond to it. Instead of focusing upon how received or embedded desires are reconstituted through technology, in a way that often pacifies them and developing a theory of reflexivity based upon analysing when the effects of reconstitution might be welcomed or rejected, Heidegger responds with the more ethereal strategy of *Gelassenheit* or 'letting be'.

Such a strategy remains forever ambivalent. There is some truth in the argument that it indicates a retreat into a romantic passivity, after Heidegger's disastrous complicity with National Socialism in the 1930s. I want to argue nonetheless that we can find hints of a more effective response to technology in Heidegger. To do this however, requires that we register the arguments against Heidegger and the criticisms of his work.

By the time he came to write on technology, Heidegger had shifted from an analytic of *Dasein*, the manner in which Being is revealed via the human subject, to a focus on the history of Being itself. Heidegger felt that his earlier work remained complicit with the metaphysics of subjectivity he was trying to counter. The later emphasis on Being tried to analyse the ways in which phenomena present themselves within an age. In this later work, the individual subject is subservient to the mode in which Being reveals the world. Retrospectively, it seems clear that with such a strategy Heidegger was also distancing himself from his political involvement with National Socialism.

The 'encounter between global technology and modern man'

The opening quotation in this chapter saw Heidegger drawing attention to the 'inner truth and greatness' of National Socialism. For him, these qualities arose especially in 'the encounter between global technology and modern man'. We can draw several things from this statement. The first concerns the link between Heidegger's commitment to the Nazis and its relationship to his early thinking on technology. The second lies in comparing Heidegger's earlier and later understandings of technology and the question of how to respond to it effectively.

Michael Zimmerman argues that much of Heidegger's commitment to National Socialism arose out of a need to respond to the onset of technological nihilism. The most influential figure arguing for the immanent arrival of the new technological age was Ernst Jünger, whose work Heidegger read and directly engaged with. While I do not intend to outline Jünger's thought in full, it remains useful to examine the technological *Gestalt* he believed would shape humanity's future, so as to understand the context of Heidegger's actions.[3] Jünger argued that the planet was in the grip of a technological evolution. He celebrated the technological overcoming of bourgeois sentimentalism and humanity, arguing that technology produced an entirely new framework through which we inhabited the world, the *Gestalt* of the worker. Like the Futurists, Jünger celebrated warfare and technological violence as aesthetic phenomena. More importantly for Heidegger, Jünger also argued that one could either submit to the new technological order, and be transformed, or be destroyed. Jünger believed that humanity's destiny lay in the technological transformation of the world into an age of nihilism. In *Der Arbeiter* he repeatedly outlines this belief:

> Our technological world is not an area of unlimited possibilities: rather it possesses an embryonic character which drives it toward a *predetermined maturity*. So it is that our world resembles a monstrous foundry ... There is no stability of forms; all forms are constantly moulded by dynamic unrest. There is no stability of means; nothing is stable outside of the rise of production curves.
>
> (Zimmerman 1990: 60, emphasis added)

Heidegger both recognised the accuracy of Jünger's description of technological transformation and rejected his affirmation of a nihilistic technological future where all previous modes of being would be swept aside. It seems clear that Heidegger turned to politics in order to forge an alternative to Jünger's predictions. According to Heidegger:

> In the year 1930 Ernst Jünger's article on 'total mobilization' had appeared ... Thinking from these writings and, still more essentially

from their foundations we thought what was coming, that is to say we attempted to counter it, as we confronted it.

(Zimmerman 1990: 36)

The attempt to counter such a technological future led, at least in part, to Heidegger's involvement with politics. Zimmerman succinctly sums up the situation: 'In the early 1930s, Heidegger believed that National Socialism offered an alternative to the technological nihilism forecast by Jünger' (Zimmerman 1990: 36). I do not, however, wish to give the impression that Heidegger was in any sense opposed to Jünger's work. On the contrary, as both Zimmerman and Wolin have shown, Heidegger relied on Jünger to a great extent for his descriptions of how technology stamped the present epoch; for the rhetoric of hardness and virility in his official National Socialist speeches, and for his belief in the transformative value of work. As Wolin points out 'it was at least partially on the basis of Jünger's theoretical apotheosis of a heroic new breed of "worker-soldiers" and "soldier-workers" that Heidegger ultimately cast in his lot with the National Socialists' (Wolin 1990: 140). The point here is that Heidegger remained profoundly convinced by much of Jünger's evaluation of the present age, but remained uncomfortable with his conclusions. Having said this, we can now return to the relationship between Heidegger, National Socialism and technology.

National Socialism's 'inner truth and greatness', according to Heidegger in 1935 lay in the encounter between 'global technology and modern man'. What is interesting about this statement is that unlike the theorisation of technology as *Gestell*, whose appropriative power transcends any political force or act of willing, Heidegger praises National Socialism for its ability to 'use' technology in a more authentic manner. Heidegger's political commitments cause him to remain tied to a muted but nevertheless clearly *instrumental* conception of technology that he will later reject. On this point, Rainer Schürmann writes:

[h]ow could a political movement be supposed capable of bearing up against global technology, unless the latter was still viewed as a force that might be stemmed and modified? Consequently in the mid-thirties technology is still seen as an inauthentic project that an authentic project could in some way redress.

(Schürmann 1987: 16)

As Schürmann points out, Heidegger argues that America and Russia represent an inauthentic relationship to technology. Heidegger writes that '[f]rom a metaphysical point of view, Russia and America are the same; the same dreary technological frenzy . . .' (Heidegger 1959: 37). By contrast Heidegger sees the potential within National Socialism to initiate a different relationship to technology. At this stage Heidegger saw political intervention as capable of determining a fundamentally different trajectory for the future than that described by Jünger.

But when Heidegger develops his mature theory of technology, he rejects the possibility that any one nation or political force could directly encounter technology and respond to it differently. As we have seen, in this later work technology is understood as a global enframing, whose manner of presencing transcends any singular attempt to respond otherwise. Indeed to imagine, as Heidegger himself had in 1935, that the manner through which technology constitutes our relation to being could be altered is, according to his later thinking, to be exposed to the greatest of dangers, misrecognising the nature of the *Gestell*.

Heidegger argues that the *Gestell* 'threatens man with the possibility that it could be denied to him to enter into a more original revealing and hence experience the call of a more primal truth' (Heidegger 1977: 333). Technology is dangerous in that it might conceal its essence as enframing, and that it 'might drive out every other possibility of revealing' (ibid.). The danger of technology, then, lies in the naive belief that we could simply manipulate it as a neutral tool in order to solve any difficulties. According to Heidegger, this can only lead into further domination, for we continue to see the world, including ourselves, as an object of manipulation, adjustable by technical and calculative thinking.

A contemporary example of such calculative thinking lies in the 'ethics' of someone like Peter Singer, who, while ambivalent about certain aspects of technological change, provides little in terms of a criteria as to why we might want to set limits to the operation of technocratic rationality. For Singer, if a technology such as IVF or cloning is able to benefit the individual, this is a sufficient criterion for allowing its operation. In a paradigmatic example of a kind of thinking that closes off other modes of revealing Singer writes in relation to the question of IVF and surrogate motherhood that 'I think the question "who is the mother?" is a pretty irrelevant question' (in Bone 1997: 1–4).[4] Such a statement ignores the deeply embedded social, cultural and biological meanings that surround the experience of motherhood. By glibly dismissing the role of motherhood as that which grants a specific form of revealing, Singer cuts himself off from a more fundamental question, namely: can the widespread operation of technologically mediated birth work to alter the meaning of what it is to have a child – in effect hollow out the ground through which childbirth has a meaning in any human sense of the word?

The real danger, then, is not the incorrect use of technology, but the denial of access to other modes of revealing which would allow us to operate within a different modality to that of the *Gestell*. Heidegger's work after the 'turn' in his thought attempts to theorise how this alternative mode of engagement might be allowed to unfold.

Whatever the problems with Heidegger's theory of technology, and whether or not his theorisation of technology represents a recoil from his political encounters of the 1930s, I want to argue that his claim that technology must be thought of as something which operates within a distinctive constitutive frame which shapes human behaviour and understanding independently of whether any particular technology is being used – is a necessary starting point for any critical engagement with the question of technology. Throughout this book we shall see how technology can be understood in this way. Whether it be Benjamin's understanding of the reconstituting function of technology in relation to aura, Lyotard's theorisation of new social bonds via computers and postmodern science, or simply the utopian fantasies of cyberculture's most prominent figures, they can all be most usefully analysed through the way that technology reconstitutes our modes of engagement, thereby initiating a new ontological framework. Heidegger recognised the reconstitutive power of technology and, though at times he conflated this into a totalising picture of the *Gestell*, he nevertheless insisted that technology could be uncoupled from this metaphysic. The remainder of this chapter looks at Heidegger's path towards this detachment of technology from the *Gestell*, examines the most common objections, and attempts to move towards an understanding of the aspects of his work that might assist in developing a critical theory of technology.

Heidegger argues that we are not powerless in the face of technology. While we cannot simply ignore or destroy technology, we can open ourselves up to its essence through *Gelassenheit*, understood as a process of releasement or letting be. According to Heidegger, this involves a strategy of waiting, of non-willing. He outlines how such a strategy could operate:

> We can use technical devices as they ought to be used and also let them alone as something that does not affect our inner and real core. We can affirm the unavoidable use of technical devices, and also deny them the right to dominate us . . . But will not saying both yes and no to technical devices make our relation to technology ambivalent and insecure? On the contrary! Our relation to technology will become wonderfully simple and relaxed. We would let technical devices enter our daily life, and at the same time leave them outside, that is, let them alone, as things which remain nothing absolute but remain dependent on something higher. I would call this comportment towards technology which expresses 'yes' and at the same time 'no' by an old word *releasement toward things*.
>
> (Heidegger 1969: 54)

While this sounds like a positive strategy, in the light of Heidegger's more overwhelming pessimism concerning the degree to which the *Gestell* has colonised our relation to being, he nonetheless fails to provide a comprehensive theoretical framework as to how such releasement might unfold across the broader social formation. At times Heidegger speaks of *Gelassenheit* as an almost mystical process. To let things be:

> is to be the one who waits, the one who attends upon the coming to presence of Being in that in thinking he guards it. Only when man, as the shepherd of Being, attends upon the truth of Being can he expect an

arrival of the destining of Being and not sink to the level of a mere wanting to know.

(Heidegger 1977: 41–42)

Heidegger clearly gestures towards the cultivation of a different response to technology that would enable us to live within the technological age, and not think of technology merely in terms of instrumental calculation. However, how we are to think and act in response to the *Gestell*, in a manner that does not correspond with willing or domination, is a question that Heidegger never effectively answers. He hoped that art, and particularly poetry would provide a space for releasement from the essence of technology. Yet taken at face value, such a strategy seems to have little use for many critics. Even sympathetic critics often seek to shed light on Heidegger's strategy of *Gelassenheit* only by reference to Eastern philosophy or mysticism (for instance Thiele 1995: Chapter 9). Instead of attempting to set limits to the operation of the *Gestell* by theorising how broader social and cultural frameworks might work to structure a different mode of relatedness to the world, Heidegger turns inward, at times solipsistically, exposing himself up to a body of criticism that is justified, even if limited (as I hope to show).

Heidegger's work after the 'turn' has been seen by some as precluding any possibility of subjective engagement with or critical appropriation of technology. As Richard Wolin argues, 'the doctrine of Being, in its oppressive omnipotence, causes the conceptual space in which freedom can be meaningfully thought to all but disappear' (Wolin 1990: 153). Indeed, on a superficial level, it does seem naive to argue for meditative thinking or 'letting things be' in the face of massive ecological destruction and the general impoverishment of life under late capitalism as it harnesses technology to its own ends. For Wolin, and many others, the problem lies with Heidegger's historical determinism. They argue that, because Heidegger claimed humanity could do nothing to alter the technological destiny into which it had been 'thrown', he could envisage no practical opposition to technology, nor any way in which technology could be used in a non-instrumental manner. Certainly, there are passages in Heidegger's writing to support these claims. For instance, in the *Discourse on Thinking*, he claims that:

[t]echnological advance will move faster and faster and *can never be stopped*. In all areas of his existence, man will be encircled ever more tightly by the forces of technology. These forces . . . since man has not made them, *have moved long since beyond his will and have outgrown his capacity for decision*.

(Heidegger 1969: 51, emphasis added)

Passages like these steer dangerously close to technological determinism. Indeed Heidegger's critics argue that in his attempt to escape the dangers of modern subjectivity, he was forced to posit a version of enframing as a destiny unchangeable by any kind of human action. Read in this light, the emphasis on passivity and submission seems to suggest a wholesale rejection of any autonomous action on the part of the human subject. Hans Blumenburg regards Heidegger's work as a retreat from the normative potentials of modernity into the quasi-theological shelter of Being. He argues that Heidegger permits no contradiction in his analysis of the modern world, thus subsuming critique and human action behind a rhetoric of submission. He claims that for Heidegger:

> The [modern] age appears as an absolute 'fact' . . . it stands, sharply circumscribed, outside of any logic, adapted to a state of erring; and in spite of its immanent pathos of destruction (or precisely on account of it) finally permits only the one attitude that is the sole option that the 'history of Being' leaves open to man: *submission*.
>
> (Blumenburg 1983: 192)

In a sense, Heidegger's critics are right to say that he did not focus sufficiently on the contradictory nature of technological enframing. Yet it would be dangerous to dismiss out of hand Heidegger's critique of autonomous action, or his theory of a historically determined technological epoch, since this suggests a return to an instrumental understanding of technology. Instead of rejecting or accepting Heidegger's critique of autonomous action, we need to consider the way in which subjective actions are determined within a broader constitutive framework. When Heidegger writes that 'everything is functioning' we can understand him to be critiquing activities that take place at an abstract level where the relevance of prior levels has diminished. To simply reinsert the category of activity into a politics of technology ignores the context through which the meanings of any activity are determined, and invites the danger of such activity degrading into mere functioning.

At this point, I want to reject some of these criticisms of Heidegger as tending to obscure a more important issue. I do believe the issue of contradiction is essential to developing an alternative to the *Gestell*. However, what matters is a more profound contradiction in the sense that it is ontological, the contradiction arising when embedded ways of being come up against a reconstitution of their contents through a more abstract setting. Such contradictions can form a basis from which to develop a theory of *reflexivity*, whereby we can engage with technology without being caught within the logic of enframing. I should again stress that contradiction and reflexivity are being used in a specific sense, revolving around the dialectical interplay of constitutive forms. I shall return to this in the final section of this chapter. However, it remains to consider which criticisms of Heidegger are indeed valid.

While it is not my intention to defend Heidegger *in toto*, it does seem that many critical responses have relied upon an almost caricatured depiction of his notion of passivity. When Heidegger seems to dismiss human action, it could be argued that this refers only to human action under the *Gestell*. In order to respond appropriately to technology, we would need to act outside the sphere of the *Gestell*. This may well be possible if we reconceive our notions of activity and passivity. We can make use of Heidegger's thinking if we try to understand his notion of passivity differently. We can distinguish between two types of passivity. The first is passivity understood in terms of a metaphysical opposition to activity, where activity is the acting upon of an external subject, and passivity being acted upon by an external force. This is how the terms are understood in the age of technology. The second mode of passivity would involve a reopening of the conditions for experience which allows different modes of relatedness with the world, outside that of the subject–object dichotomy carried to an extreme under the *Gestell*. It is dangerous simply to dismiss Heidegger, by invoking categories of action and politics because this ignores the way these categories are themselves embedded within a broader framework. Indeed, it is only through a more sophisticated understanding of how such frameworks determine our modes of relatedness that we can respond more productively to the spirit that motivates Heidegger's notion of letting things be. I shall return to this issue in the final section of this chapter.

However, many critics have questioned whether a second approach is at all possible given Heidegger's ethical shortcomings, both at the personal and (more contentiously) at the theoretical level. It is to this question that I now turn.

The status of the Other in Heidegger's thinking of the technological question

Even if we accept the potential of Heidegger's call for a passive response to technology, there still remain fundamental flaws in his thinking. These lie in two related areas: first, in Heidegger's tendency to conflate all activity under the rule of enframing; second in his apparent inability to respect or sufficiently account for the ontological status of the Other. Both problems can be analysed by reference to what is perhaps Heidegger's most scandalous pronouncement. Interestingly, it occurs during one of the rare moments where he refers to the Holocaust:

> [a]griculture is now a motorised food industry – in essence the same as the manufacturing of corpses in gas chambers and the extermination camps, the same as the blockading and starving of nations, the same as the manufacture of atom bombs.[5]
>
> (Heidegger in Neske and Kettering 1990: xxix–xxx)

The first flaw is perhaps the most obvious. While gesturing at times toward the possibility of a 'free' relationship to technology, Heidegger cannot provide a comprehensive criterion for determining the ethical and political considerations that would sustain such freedom. In pronouncements such as this he offers a view of technology that allows little chance for its constructive employment. Ethically, it is clearly possible to distinguish between mechanised food production and the ovens in Nazi death camps. Heidegger in this statement reveals what he earlier described as the danger of the *Gestell*, that it would obscure every other possible mode of revealing. The lack of distinction between these phenomena would seem to indict Heidegger on his own terms. He remains within a one-dimensional framework that cannot account for specific political movements, national ideologies and the needs of the tangible Other.

Yet there can be no doubt that technology, understood as enabling a more constitutively abstract mode of engagement with the world, is a necessary precondition for the phenomena Heidegger listed. To reject his comment out of hand would be to cut us off from a crucial understanding of how processes of abstraction can lead to an impoverishment of any moral imperative. Arne Johan Vetlesen situates the possibility of intersubjective ethics in relation to the degree of proximity between subjects, arguing, in effect, that 'moral capacity is fostered within a small-scale setting' (Vetlesen 1993: 374), and conversely, that actions performed within a social context of abstraction operate in a space where violence and domination can be committed more easily and effectively. Drawing upon the work of Zygmunt Bauman, and citing the findings of psychologist Stanley Milgram,[6] as well as accounts of the Holocaust, Vetlesen argues that processes of abstraction cause the framework whereby we can develop the capacity to respect the existence of an Other to recede. Sounding in one sense very much like Heidegger, Vetlesen comments on the processes that enabled the actual implementation of the Holocaust:

> it was absolutely instrumental for the planners of the Holocaust to subject the chosen target, the Jews, to various processes of *abstraction:* Only when rendered abstract, when turned into an invisible, *versachlichte* far-off 'target' at the remote receiving end could the goal of killing millions be obtained – this precisely *because* the undertaking had lost its character as downright murder and . . . [was] redefined and disguised as a technical task no different in nature from every other technical task, though admittedly a particularly 'tough' and 'complicated' one.
>
> (Vetlesen 1993: 344)

I recognise that this is a delicate issue. I do not wish my qualification of Heidegger's comments to serve as an excuse for his behaviour. Nor do I wish to lessen the significance of the Holocaust or elide the fact that it was a specifically German phenomenon, or that many other factors were involved and could distinguish the Holocaust from the phenomenon of atom bombs or agribusiness. On the other hand, I think it is vital to register that on one level Heidegger is correct; that all of these activities, each of which is devastating (historically or potentially) on a massive and heretofore unprecedented scale, is possible because of the *essence* of technology which through a process of abstraction allows subjective actions to be removed from their concrete responsibility.

In this sense, Heidegger's comment does suggest how material processes of abstraction form an essential precondition for a sensibility that is able to produce the Holocaust, the starvation of nations through agribusiness, or the construction of weaponry which could annihilate all forms of life. It is worth repeating that technology provides both the material capacity to carry out these terrible things, and also the constitutive framework where, because the other is regarded merely as standing-reserve, it becomes easier to do violence to them. While acknowledging that Heidegger's remark was used to lessen the significance of Nazi atrocities and his own implication therein, Leslie Thiele writes that:

> Nazism was a technologically driven enterprise in which concern for the world was denied so that the lust for domination might run rampant. Agribusiness, in turn, is a technologically driven enterprise in which concern for the world is subordinated so that the lust for profit may run rampant. In each case, technological mastery overwhelms the needful relations human beings might establish with the earth and with each other.
> (Thiele 1995: 144)

Heidegger's insensitivity consists precisely in the fact that his analysis remains locked within this more abstract level: *in theory he carries out the very process of levelling that technology (as Gestell) carries out in practice*. The need remains to distinguish between modern agriculture and the Holocaust. As too does the need to establish 'needful relations . . . with the earth and with each other'. The question is how can this be achieved and, more specifically, can it be achieved at all within the parameters of Heidegger's thought? One approach has been to correct the perceived ethical lack in Heidegger's thought through a thinking of the concrete Other in its various manifestations, as victim, or more generally as an embodied and finite being.

For Heidegger, the most dangerous thing about technology is its capacity to disrupt things from their essential nature, or authentic coming to presence. For some critics, this leads to disturbing consequences. John Caputo argues that Heidegger's thought is governed by a type of 'phainesthetics', which stress the experience of Being as the self-showing of the *phainomenon*, a primal presencing that comes toward us and concerns us. The thinker must be able to respond to this primal call and allow things to come to presence. Being means appearing, but it is the *manner* in which things appear that is essential. Caputo argues that, in attempting to recover this presencing, in remaining attentive to the call, much is forgotten. In a sense, Being is elevated above life itself, literally, Caputo argues, because it ignores the sufferings of the victim. In valorising Being on phainesthetic grounds, Heidegger can only understand experience by reference to world and thing. Consequently, the thinking of *beings* in their *corporeal* actuality, in their ability to feel, suffer, live or die, in particular in their presence as victims, is elided. The call of Being is deaf to the cry of the victim, to the knowledge of its suffering. Ultimately, '[t]he assault upon the earth which turns the soil into an object of agricultural engineering is more primordial than the ravages of hunger, of ravaged bodies' (Thiele

1995: 268). An overriding concern with how things come to presence also fosters the ethical blindspot which then allows Heidegger to equate motorised food production with production of the victims of the gas chambers.

For Caputo, 'Thinking' about Being in this manner involves prioritising *manner* over *matter*. The manner in which things are allowed to show themselves, gas chambers, the mechanised food industry, is deemed more important than the matter itself, murder, starvation. Consequently, the victim remains unthought in the history of Being. Caputo remarks that Heidegger's world is:

> a world in which a wholly *other* kind of responsiveness and responsibility has been silenced, a responsibility to those who live and die, to those who are embodied, who suffer or are in pain, who grow old or infirm – above all to innocent victims. The thinker leaves no room at all for the victim in the history of Being's self showing.
>
> (Thiele 1995: 277)

The question remains as to how we can reinsert the victim (or even the existence of the concrete Other) into the search for the meaning of Being. Caputo's distinction between manner and matter reveals a disturbing absence in Heidegger's thought. But while it is important to focus on the problem of 'matter', I want to argue that such an approach is limited. Indeed, I want to say that we can more effectively theorise the status of the victim by returning to the framework that Caputo names as 'manner'. Of course this is to move away from Heidegger and to understand the question of 'manner' as a question of constitutive levels. First, however, it is necessary to point out the benefits and limits of theorising the question of 'matter'. For some critics this involves reinserting a fuller notion of corporeality into Heidegger's thought.

Heidegger does not specifically focus on the constitutive role of the body in his work on technology, remarking that 'this "bodily nature" hides a whole problematic of its own' (Heidegger 1962: 143). As we have seen, he elides the consideration of the victim's body. An emphasis on corporeality may provide a way beyond Heidegger. At the end of his remarkable study, Zimmerman suggests that there may be an alternative way to understand the disclosive power of *techne* than through reference to the saving power of art or *Gelassenheit*. This involves understanding how the human being is disclosed as a corporeal subject. Zimmerman suggests that, instead of the vain hope that poetry will release us from technological enframing, we might instead develop 'a new *techne* of the body', through which in 'the art of disclosing, encountering and affirming our own mortality, corporeality and suffering ... we could learn to "let ourselves be"' (Zimmerman 1990: 247). For Zimmerman, then, an understanding of the importance of embodied practices may enable a more effective resistance to *Gestell*. There is no doubt that the technological imperative is often underwritten by a desire to evade the reality of the embodied, finite being, by attempting to transcend it technologically.

We may remember that Heidegger despaired at the technological age because it enabled us to escape the question of our mortality, writing that: 'The self assertion of technological objectification is the constant negation of death' (Heidegger 1971: 125). Heidegger argued that only through acknowledging their finitude could humans free themselves of the technological will to power. By coming to terms with the significance of our finitude, he claimed that we could free ourselves from the voracious appetite of groundless subjectivity that comes to the fore in the modern age. There is little doubt that technologically-driven activity coincides with a desire to evade death, and that this is linked with the desire to transcend the lived body, for this is also a body that decays and dies. For example, Bob Ettinger, author of *The Prospect of Immortality*, equates the tragedies of 'the human condition' with the limitations of nature and the inherent mortality of humankind. He writes that people have 'cheap bodies, erratic emotions, and feeble mentalities . . . to be born human is an affliction, it shouldn't happen to a dog'.[7] The answer is to change the 'flaws with the engineering', essentially to make us immortal. As Ed Regis, paraphrases the argument:

> It was no consolation that all of these shortcomings were quite understandable in evolutionary terms. Mankind, after all was a product of nature, and nature worked not by intelligent planning and conscious design but by the worst kind of trial and error blundering: try this, try that, and see what worked out. Mostly things *didn't* work out, as was clear from the fact that the overwhelming majority of the species that had ever evolved became extinct soon afterward. *So it was no surprise that human beings were as botched up as they in fact were.*
> (Regis 1990: 145, emphasis added)

The desire to evade death, the belief that humanity's problems lie in its mortality, the denigration of corporeality, and the belief that humanity can be improved by tinkering with the basic engineering, all intertwine in a way that confirms Heidegger's analysis. Yet, in trying to emphasise the importance of finitude as a means to provide a different response to technology, Heidegger did not explicitly concern himself with the phenomenon of the body. While Heidegger claimed that coming to terms with death could provide the means to an authentic relation to being, he remained at times oblivious to the deaths of others. While he would have abhorred the desire to flee the body technologically, his indifference to the body of the Other that dies or suffers indicates a thoughtlessness of a different kind. How, then, might we establish an openness to the embodied Other?

One way is to tackle the question of embodiment through thinking corporeality as a question of tactility, a mode of engagement that some writers claim resists both objectification and the division of self and other. I will examine how the theme of embodied understanding has been tackled in relation to Heidegger, then argue that such a strategy needs to be linked to the broader issue of constitutive framing in order to allow for a more ethical relation to the concrete Other.

Tactility and the irreducibility of the Other

Heidegger claimed that the technological age caused the transmutation of the thing into a mere object. Things stand against us in some way, they maintain their essential nature. Proper use of a thing means that it remains irreducible to us, it retains its otherness. Heidegger writes that '[p]roper use does not debase what is being used – on the contrary, use is determined and defined by leaving the used thing in its essential nature' (Heidegger 1968: 187).

Focusing on the role of embodied understanding may help to resist this reduction of things to objects, the obliteration of otherness as the aggressive subject remakes the world in its own image. In *The Body's Recollection of Being*, David Levin (1995) focuses on the role of the body in human understanding. His work allows us to 'flesh out' the hints, gaps, and silences in Heidegger's work and suggests a way to respond to the nihilism that threatens us in the technological age. Levin claims that an embodied awareness provides a means for resisting the reduction of things to objects. He argues that tactile understanding preserves the alterity Heidegger felt was essential in resisting the domination that arises through the subject-based conception of the world. Tactility allows us to use things, yet let them remain free of our use. Referring to Heidegger's analysis of equipment in *Being and Time*, Levin argues that tools 'are certainly useful; but they are also manifestly beautiful – and not only when they are left to stand in their own intactness, but even when they are actually being used, if handled by hands receptive to the moving beauty of their presence' (ibid.) He concludes that tactility might place us within a framework more receptive to the existence of the other, and more responsible for our own actions.

Heidegger convincingly argued that technology increased the subject–object polarity, until the point where the alterity of the object itself seemed to disappear. This development corresponds with the devaluation of embodiment: we lose touch with Being. Levin claims that the importance of embodied understanding lies in the fact that it decentres the subject from his or her ego-logical framework. For example, the simple act of touching invites a reciprocity to be touched. This opens up a field of awareness that transcends the division of subject and object. The emphasis on embodied knowledge may also allow us to focus on the importance of alterity, and perhaps allow for a more concrete sense of the other to emerge, opened up through engaging with non-technical criteria.

The focus on the body as a means of grounding and affirming our sense of finitude, or on embodied tactility as providing a means to open oneself to the other, rather than dominate it, is suggestive of ways to set limits to the operation of technological abstraction through the encouragement of alternative experiential paradigms. But in themselves these are not sufficient. We need to ask what would sustain the *meanings* of the body, in the sense that would concur with the suggestions of Zimmerman and Levin. The process is by no means an automatic one. For instance, in recent years we have seen a renewed focus on the body, both in cultural theory and in popular culture. However, this return does not automatically generate the sense of openness that we might wish for. The meanings of embodiment can themselves be reconstituted within a different framework. All too often the meanings of the body are produced through the *Gestell* – understood entirely within the context of a postmodern hyperindividualisation where the body becomes a means to subjective exploration and individual empowerment. This type of embodied understanding closes off an approach to the Other as it is subsumed within the more abstract modality of subject constitution. Alternatively, the tactile body is itself caught up in a process of abstraction, its pleasures harnessed and reconstituted as commodities or its senses engaging with the other only at a distance, through technologically mediated forms of touch and sexuality. In the absence of a binding framework such as the face-to-face, the link between embodied understanding and an ethical openness that might supplement Heidegger's work remains tenuous. To say this is not to dismiss the focus on the body as an ethical site of resistance to technological domination. Rather I wish to point out the limitations of focusing on the body without considering the constitutive framework through which the meanings of embodiment might unfold. Embodiment is always a contextual process, inscribed in practices across the social realm. As Paul Connerton reminds us; bodily practices 'cannot be reduced to a sign that exists on a separate level outside the sphere of the body's acts' (Connerton 1989: 95). I want to argue that we might more productively engage with this issue by thinking the question of the body and the other through a return to the exploration of constitutive frameworks (or in Caputo's terms, manner).

At this stage it might be useful to recap briefly on what has occurred so far. I have argued that Heidegger's understanding of technology as something which constructs a specific constitutive framework, provides an essential starting point for any work towards a critical theory of technology. His description of how technology works as a process of enframing, which facilitates a metaphysics of autonomous subjectivity remains convincing and relevant. However, there remain problems with his work – in particular, problems of overdetermination and of an inability to theorise effectively a mode of resistance to the process of enframing. There is also the problem of ethical differentiation where the fate of beings is subsumed under the question of being. Heidegger's strategy of *Gelassenheit* seems unable to stand up effectively to the totality of these criticisms. Partial solutions have been suggested, through the notion of supplementing Heidegger's work on technology with approaches that crystallise around this question of matter, namely theorising and accounting for the embodied being. I have suggested that this approach, while useful, is limited, and that a more productive approach might be to return to the question of *manner*, or to use my own idiom, constitutive framing.

How we can use Heidegger to say a 'yes and a 'no' to technology

The final section of this chapter suggests that theorising the manner through which things appear, a central preoccupation in Heidegger's work, remains of vital importance. More importantly, we can begin to outline a theory of 'manner', as the constitutive framework which grounds and lends meaning to actions and values, so as to include the consideration and respect of the concrete Other, in this respect going beyond Heidegger. Such a project would emphasise the importance of a *plurality* of modes of revealing, but also of attempting to understand the relationship between these modes. Such understanding can be gained through concerning ourselves with the context through which any revealing takes place, and the degree of abstraction at which it occurs. As we have seen, technology as a mode of revealing allows things to be reconstituted at a greater degree of abstraction. It is through understanding this process, and then reflexively determining whether we might welcome the reconstitution of our modes of engagement with the world (through considering the ontological contradictions that may occur) or rather prefer to preserve prior modes of revealing, that we can effectively learn how to say both yes and no to technology.

But we must first turn to the question of a plurality of modes of revealing. In an article which makes a number of perceptive criticisms of Heidegger's 'The Question Concerning Technology', Richard Bernstein (1991) identifies two silences in Heidegger, the first of which is the notorious silence concerning the Holocaust. The second concerns the elision of *phronesis*, practical thinking or politics, in Heidegger's discussion of technology. Here, Heidegger relies on Aristotle's discussion in the *Nichomachean Ethics* of how *alethia* (truth as a revealing) may appear. Bernstein notes how Heidegger remains silent with regard to this central category of revealing:

> [t]hroughout his essay Heidegger speaks as if there were a plurality of modes of revealing, but he only explicitly considers two modes, poiésis (bringing forth) and *Gestell* (challenging-forth) ... the entire rhetorical construction of the 'Question Concerning Technology' seduces us into thinking that the only alternative to the threatening danger of *Gestell* is poiésis.
>
> (Bernstein 1991: 120–121)

Bernstein's criticism touches upon a critical issue, that of the need for a plurality of modes of revealing. I am making a similar claim when I argue for a conception of the social as composed of the intersection of various levels. I want to take up this issue here, and suggest that we can expand upon the strategy of *Gelassenheit*, broadly understood as a means of setting limits to the operation of technological metaphysics, by thinking through the notion of a plurality of modes of revealing and the relation between them. It is the relationship between these modes that strikes me as a vitally important issue, and one upon which Bernstein himself is silent. In Chapter 1, I argued that less abstract constitutional forms, ground and intersect ways of being and acting that are more abstractly constituted. To establish a dialectic between these more or less abstract forms would allow a means of saying 'yes' to technology. It is time to return to this claim, through a discussion of some of the more neglected aspects of Heidegger.

Despite Bernstein's criticism of 'The Question Concerning Technology', elsewhere Heidegger does discuss other modes of revealing rather than simply hovering between the stark polarities of *Gestell* and poetic revealing (Theile 1995).[8] In 'The Origin of the Work of Art' Heidegger writes:

> But how does truth happen? We answer: it happens in a few essential ways. One of those ways in which truth happens is the work-being of the work . . . Beauty is one way in which truth occurs as unconcealedness . . . Another way in which truth occurs is the act that founds a political state . . . still another way which truth becomes is the thinkers questioning, which as the thinking of Being, names Being in its question worthiness.
>
> (Heidegger 1971: 55–56)

Here Heidegger notes that politics can be regarded as a mode of revealing. Given his political career, we may well wish to shy away from any discussion of politics. But what is the nature of this mode? How does it relate to other human activities? I want to suggest that Heidegger conceives of politics as a mediating constitutive *form* when he writes that 'to each mode of founding there corresponds a mode of preserving' (Heidegger 1971: 35).[9] What, however, is being preserved? It is here that Heidegger remarks that this mode of revealing is predicated upon a kind of work. He notes that:

> [u]nconcealment occurs only when it is achieved by work . . . the work of the word in poetry, the work of the stone in temple and statue, the work of the word in thought, the work of the polis as the historical place in which all this is grounded and preserved.
>
> (Heidegger 1959: 191)

This is a form of work that for Heidegger, is not governed by technological enframing – it is not merely 'functioning'. In theory, it is a work that correlates to the non-metaphysical mode of passivity, a work that at the same time also preserves. It is interesting that Heidegger regards the polis as neither a formal city nor a state, but rather as a site or place that grounds and preserves a more originary relation. This indicates first, the social aspect of this mode of revealing and secondly, the nature of the polis as an ontologically constitutive form – not for the playing out of a particular historical destiny, but as a site where social activity can unfold and preserve a certain relation to being. Heidegger writes that the polis is a mode that grounds and preserves. We can extrapolate and claim that the polis is a mode that preserves the kind of relations that open us to the tangibility of the Other. In other words, it acts as a *mediating* form between modes which reveal and bind our relation to the concrete Other, and the more abstract forms of revealing made available through technology. This would constitute the 'work of the polis' as a mediating form that preserves a mode of revealing outside of the *Gestell*. Using the terms outlined in the previous chapter, Heidegger's discussion of the polis as a specific mode of revealing that preserves a prior relation, corresponds with the manner in which more abstract modes of integration draw upon and mediate relations carried out within the face-to-face level.

The emphasis on the polis as a form or site, rather than a place or mode with any specific content helps to explain Heidegger's view of how politics is to respond to technology. He remarks that: '[f]or me today it is a decisive question as to how any political system – and which one – can be adapted to an epoch of technicity ... I am not convinced that it is democracy' (Heidegger 1981: 55). This statement can be read in two ways. It can, of course, be read as evidence of Heidegger's defeatism, of his inability to provide any real response to technology. On the other hand, it can also be read as a warning not to privilege questions of political content over their formal conditions of possibility. In some ways, Heidegger shares with Jean Baudrillard a suspicion of democracy calculated around an abstract technocratic thinking (for Baudrillard, based around the simulations of media, polls and surveys). This would be an abstracted democracy, lifted out from the ontological form or site that once preserved its particular mode of revealing, a democracy that could only unfold within the terms of the *Gestell*.

Heidegger wrote that the danger of technology lay in the way the one-dimensional framework of *Gestell* denied the existence of other modes of relatedness. Despite his tendency to overstate the degree to which technology has hollowed out or concealed other modes of revealing, he maintained that we could say 'a yes and a no' to the use of technology. We have seen how Heidegger notes the existence of contradictions, such as when the subject is forced to take itself as standing-reserve, and how he fails to go beyond the acknowledgment of this contradiction. We have examined the need to conserve other modes. How does this allow us to say yes and no to technology? To say a yes *and* a no to technology suggests the ability to adopt a certain reflexivity. This would require an understanding of the way technology works to reconstitute human modes of relatedness to the world. In some cases this may be desirable; though Heidegger was much more eloquent in pointing to cases where it was not. What enables the grounds for such reflexivity is an understanding of the interplay of different modes of revealing. I have gestured towards an alternative understanding of Heidegger's conception of politics, as that which preserves a specific mode of relatedness. It is here that I also want to emphasise the importance, for Heidegger of Being-with-Others as providing an essential ground for an authentic relation to Being (and for providing a way outside the *Gestell*).

While Heidegger's later work tends to situate the question of alternative modes of revealing within a framework that disregards any social framing of the issue, there are aspects of his earlier work which suggest the fundamental importance of the *social* aspects of our being. In *Being and Time*, Heidegger writes that 'Knowing oneself [Sichkennen] is grounded in Being-with, which understands *primordially*. It operates proximally in accordance with the kind of Being which is *closest* to us – Being-in-the-world as Being-with' (Heidegger 1962: 161, my emphasis).

For Heidegger, knowledge of oneself is grounded in Being-with-Others. He emphasises the importance of this Being-with by describing this social being as 'primordial' and 'closest-to-us'. Heidegger goes so far, at one point, as to imply that an authentic self is *only* possible through the relation of Being-with. He writes that 'Being-with-Others belongs to the Being of Dasein, which is an issue for Dasein in its very Being. Thus as Being-with, *Dasein "is" essentially for the sake of Others* . . .' (Heidegger 1962: 160, emphasis added).

Heidegger's description of this relation resembles the face-to-face modality of social integration. It is also significant that his later work makes it clear that more abstract relations are bound by this primordial relation. Heidegger's emphasis on the shared nature of being also extends to his later discussions of language. If, as I have already argued in the introduction, techniques such as writing allow for a more abstract mode of social interaction, in that the presence of the other is no longer structurally necessary, Heidegger continually asserts that all forms of authentic language are grounded in a more primordial social dimension. For Heidegger, language necessarily presupposes a prior disclosure. While this has an obvious ontological sense in the disclosure of truth as *alethia*, I want to argue that it also has an ontic dimension, that this prior disclosure is a primary social bond framed through co-presence. On this point Heidegger writes that 'the very sense of any discourse is to discourse to others and with others' (Heidegger 1985: 263). It is clear that Heidegger is emphasising the binding nature of this relation. One of the dilemmas surrounding technology, especially in our own time where communications technologies have made it possible to communicate in the absence of the tangible other, is how to maintain this primordial social relation.

I have already indicated, both here and in the introduction, how technology operates to reconstitute this most primordial relation within a more abstract setting. This can lead to the violence of the *Gestell*, where we lose our ethical capacity to respect the other because they no longer appear in a sufficiently tangible relation to ourselves. On the other hand, such abstraction can lead to a process of reaching out and becoming aware of distant others, which enables us to break free of parochialism and move towards the establishment of a globalised ethic. The question remains as to how we can preserve the primordial sense of Being-with within a context of technological mediation. It is worth returning to Heidegger's claim that the essence of technology is the *cul-*

mination of an older metaphysic. By implication, this means that technology creates the conditions through which this historically unfolding metaphysic can expand into a global type of enframing. Historically, the global entrenchment of such a metaphysics has been prevented by the various 'physical' and 'natural' limits that have resisted the transformation of the world into standing-reserve. In the twentieth century, however, many of these can be technologically overcome. Heidegger repeatedly drew attention to the disastrous implications of simply accepting the transcendence of these limits, and for this reason alone his work remains vital. On the other hand, Heidegger never fully resolved the issue of how to engage with technology differently. He clearly recognised the potential perils of a more constitutively abstract mode of engagement with the world. However, his analysis of this mode under the rubric of the *Gestell* is unable to indicate how we might forge a different relationship to it.

Throughout this book, I suggest that this may be possible through preserving, at a cultural level, aspects of these historically prior constitutive forms. Ontologically, these modes serve to limit the degree to which we might gravitate towards the kind of being that unfolds through the *Gestell*. By recognising that certain meanings are grounded in more concrete forms of being we would also recognise the degree to which their technological reconstitution within a more abstract setting might threaten their very existence within a retrievable semantic horizon. To do this would be to think through productively the potential ontological contradictions that arise from any reconstitutive process.

Using various threads drawn from Heidegger, I have argued for the importance of preserving a plurality of modes of relatedness, and suggested that we might profit by noting the contradictions that result when these modes intersect with one another. However, the question remains as to how the interplay between these modes can appropriately shape and limit human activity and thought so as to be able to engage fully with the question of technology. I do not think we can sufficiently answer these questions by remaining within the terms of Heidegger's work. We can perhaps move towards an answer by examining the work of Walter Benjamin, whose discussion of technology at times resembles Heidegger. Benjamin also provides important clues as to how we might harness the *radical potential* of technology, an orientation that Heidegger gestured to, but did not expand upon.

3 Walter Benjamin and technology
Social form and the recovery of aura

Like much of his other work, Walter Benjamin's consideration of technology is scattered, fragmented and perhaps even contradictory. Yet this tension in his writing is often ignored, especially by commentators attempting to co-opt his thought for the potentials they envisage within postmodernity. Benjamin is often seen, rather simplistically, as an advocate of the radical possibilities inherent in technology. Bound up with his treatment of technology is Benjamin's account of the historical loss of experience. It is possible to trace a dual response to this perceived loss: on the one hand, there is an attempt to win back these complex modes of experience; on the other hand, Benjamin at times seems completely to abandon older, more 'auratic' forms of experience and to replace them by a 'new pattern of perception that is low in experiential content, but for that reason precise with regard to the matter and unimpassioned' (Honneth 1993: 91). Yet if more sophisticated readings of Benjamin are able to reveal this nuanced response to new technologies, I want to argue that it is necessary to go beyond simply registering increasingly subtle forms of ambivalence in his work.

This chapter will indicate, first, that Benjamin's speculations on technology go beyond simple advocacy; in many cases he seemed to lament the changes it brought about. In addition, I hope to show that the one-sided reading of Benjamin actually deradicalises the possibilities he saw in the relations between technology, culture and society, a conclusion that contradicts those who too easily situate Benjamin as the radical forerunner to postmodernity, the celebrant of technologically produced mass culture and the wilful proponent of the destruction of the 'aura' of tradition. Nevertheless, even if it can be shown that Benjamin articulated a radical potential in auratic experience, this in itself does not take us very far. The forms of experience deemed auratic would also need an account of the social forms which frame the significance of these experiences. The argument being made in this chapter extends the claim made in Chapter 1, that the 'meanings' made possible through technology are ultimately integral with the social form in which they are carried. If technology can radically reconfigure the constitutive level which originally carried that meaning, there remains a need to develop a framework which could theorise the contradictions involved in this process. For instance, (taking the most well known of Benjamin's claims) if technology is able to 'destroy' the more hegemonic currents of tradition, does this automatically lead to a more equitable social practice? In revealing the ambivalence inherent in this destruction, however, I want to connect this back to the broader question of an ethics of technology. Before this is possible, however, Benjamin's general position on technology must be staked out.

Technology manifests itself in numerous ways in Benjamin's work. Most famously, it occurs through the theme of mechanical reproduction and the related issues of aesthetics and politics. Second, he is concerned with the transformations of experience that occur in the shift to modernity. Here technology, when linked with capitalism, facilitates the decline of meaningful experience, in the workplace, in the mode of communication, and in the breakdown of community. Often this decline is discussed under the theme of 'shock'. Third, technology functions by creating compensatory forms of experience that enable the subject to cope with such experiential shifts. The rise of phantasmagoric forms of experience, often constituted through technologically rendered environments, will be discussed in this context. Finally, there are Benjamin's fragmentary comments on the relationship between technology and nature. It is in these comments, however, that we are led to rethink the notion of auratic experience. This chapter will argue that it is possible to find a general theory of technology in Benjamin's work that goes beyond the issue of technology's relation to culture and aesthetics. In order to do so, it is necessary to contextualise his work in relation to the theories of technology that developed around him, in particular, the work of Ernst Jünger. While not the only writer Benjamin had in mind when developing his thought on technology (one thinks of the Futurists), Jünger's work was an important point of departure nonetheless. If it is not within the scope of this chapter to consider all of the 'reactionary modernists',[1] Jünger's work can, in many ways be read as emblematic of the trends that Benjamin opposed.

Benjamin did at times risk a historical gamble on the radical possibilities of technology itself. Clearly, we can say now that this gamble has largely failed. However, this does not negate his work on technology. On the contrary, I want to show why Benjamin ought not to be read in this one-sided manner, nor read as setting up an irreconcilable opposition between progressivism and nostalgia. Instead, I wish to elaborate upon a point made by Irving Wohlfarth (1978: 60) in his discussion of Benjamin's 'destructive character'. Wohlfarth suggests that Benjamin's notion of destruction ought to be approached dialectically, in that a distinction must be made between authentic traces and their false substitutes. For Benjamin, only the latter need be destroyed. This distinction can usefully be applied to Benjamin's elaborations concerning technology, allowing us to unpack the notion of aura so that several varieties of aura become apparent. When Benjamin celebrates the destruction of aura, it is a certain type of aura, functioning as an inauthentic trace, that concerns him. On the other hand, when he seems to contradict himself by lamenting the destruction of aura, it is the form of aura *ontologically prior* to its inauthentic trace that he is discussing. By ontologically prior, I mean auratic experiences carried within less abstract constitutive levels. In Chapter 1, I pointed out how these prior levels remain ontologically significant in so far as social and cultural meanings remain embedded within them. The final question then becomes: what relation to technology can be developed that will help destroy any inauthentic trace, while at the same time allowing access to a more primordial form of auratic experience that can both fuel revolutionary desire and initiate an ethical framework? I suggest that the answer lies in the nature of the constitutive form that structured this prior mode of auratic experience and in establishing a relationship to technology that can reflexively engage with this form.

The technological destruction of aura

In essays such as 'The work of art in the age of mechanical reproduction' and 'Experience and poverty', Benjamin celebrates the capacity of technology to shatter humanity's ties with both myth, and what he takes to be the more disempowering facets of traditional aesthetics. These aspects are bound up in the notion of 'aura'. Benjamin traces the function of aura through history and argues that it is bound up with notions of authority, singularity, ritual and cult. When endowed with aura, objects have a fetishistic quality, which enables them to have a mystifying authority. Aura comes to function as the cultural legitimation of traditional social functions. It is the aura's spatio-temporal singularity that lends authenticity to the objects bound up with it. This in turn bestows authority on the traditions these objects come to stand for. Benjamin claims that 'the uniqueness of a work of art . . . is inseparable from its being embedded in the fabric of tradition' (Benjamin, W. 1968: 223). Therefore, anything that disentangles us from tradition is looked upon favourably. In 'Experience and poverty', Benjamin states this claim more strongly, advocating a new start that embraces the contemporary impoverishment of experience; to 'begin from the beginning, make a fresh start, make do with little' (Benjamin, W. in McCole 1993: 156). Here he embraces the crisis in experience, since it facilitates a break from tradition. A radical antihumanism is posited, where a 'new positive concept of barbarism' (ibid.) will clear out the baggage of tradition and create a space for renewal. The technological destruction of aura clearly plays a major role in furthering this positive barbarism.

Mechanical reproduction works to eradicate the notions of singularity and uniqueness that bind the auratic object to tradition. Now, every work of art can potentially be produced *en masse*. Such massive reproduction alters the status of the original, as Benjamin notes: 'to ask for the authentic print makes no sense'. The loss of singularity leads to the loss of the authenticity of the artwork and, in turn, of the cultural tradition to which it belonged. It is not the mere possibility of reproduction that produces these changes, but rather, the historical moment when 'mechanical reproducibility became constitutive for art, when, therefore, massive reproduction became the law of form for the work itself' (Paedzoldt 1977: 47). Reproduction technology also facilitates the shrinking of distance, the other factor integral to the production of aura. The reproduction of art means that the reproduced object can be received away from the site of the original. This change allows 'the original to meet the beholder halfway'. Despite the fact that the original work of art remains untouched, 'the quality of its presence, its "here and now" is always depreciated' (Benjamin, W. 1968: 221). Thus the ontological status of the work is irrevocably altered.

The liquidation of aura and tradition is regarded by Benjamin as a largely positive development. He notes that mechanical reproducibility detaches the object from its original authoritative context, liquidating the traditional value of the cultural heritage. The object, once detached in this way, is 'reactivated', leaving the subject free to engage with it in a new light. It is important to emphasise that Benjamin envisages a new, non-auratic mode of reception developing, not as an abstract programme coming from outside, but as an objective, immanent development that arises through the proliferation of technologies of reproduction. Moreover, through multiple reproductions, works of art simply become more available to the masses. Thus a more democratic approach to art becomes possible, as art becomes available to those outside the ruling classes. Finally, mechanical reproduction generates a new mode of perception, which will, according to Benjamin, generate in the masses a more critical stance towards experience. Film and photography, for instance, allow for a more critical and testing attitude towards reality. This new mode of perception occurs partly through identifying with the technical equipment, for example, in film where, according to Benjamin the audience assumes the camera's active and constructive position rather than one of contemplative passivity. Benjamin claims that a more intensive exploration of the visual environment leads to a 'deepening of apperception' (Benjamin, W. 1968: 235). The fostering of a self-conscious critical and testing attitude made possible though the technological extension of human perceptions is itself an ambivalent process, as Heidegger's description of technological 'functioning' and Virilio's notion of a 'constitutive dispersal' (see Chapter 6) attest to. However, by analysing the relationship between experiences and their broader framing contexts, I hope to establish a more productive reading of this ambivalence.

Technology works objectively to destroy aura, but Benjamin also combines this with the historical change evident in the attitudes of the masses. According to Benjamin, the modern desire of the masses is to bring things nearer, i.e. to shrink the spatio-temporal frame through which we understand the world. Many commentators have claimed Benjamin as one of the first to envisage the potentials in technologically-produced culture. In one sense they are right, for Benjamin certainly did attempt to outline an *objectively* positive, non-instrumental telos in the rise of new technologies. I want to argue that he did this in order to show that the right-wing aestheticisation of technology was, in his terms, 'unnatural'. However, there remain many problems inherent in this position. Indeed, even in the 1936 essay, Benjamin was more ambivalent about technological reproduction than is initially apparent. However, it is necessary to outline a few of the problems involved in the destruction of aura and tradition, before we consider these broader issues.

Problems with Benjamin's theory of auratic destruction

There are several problems with Benjamin's position as outlined in 'The work of art'. These include the abolition of a critical distance, the denial of aesthetic experience to the masses, and a reduction of experience to its most impoverished form. Benjamin sees technology as assisting the masses' desire to bring things close, to 'overcome the uniqueness of reality by accepting its reproduction'. In other words, technology aids the masses in their desire to desacralise reality. Benjamin suggests that this desire is primarily 'the mark of a perception whose "sense of the universal equality of things" has increased' (Benjamin, W. 1968: 223). Given this, there is a need to ask whether these developments, which Benjamin sees as objectively unfolding through technology, are indeed progressive? As I have already pointed out, the meanings and effects of technological mediation are dependent upon a complex relationship between constitutive forms and their determination of ways of being and acting in the world. To bring things closer, to universalise our categories of experience, does not necessarily lead to a favourable outcome. We have seen how for Heidegger, bringing things close furthers the process of technological domination, rather than allowing for new freedoms. This relates to the second problem. Many writers have pointed out how the emergent modes of being made possible through technology have been harnessed for less desirable purposes.

For instance, critics have been troubled by what they regard as Benjamin's uncritical celebration of auratic destruction. Christopher Wise comments on the masses' desire to 'bring things closer', noting that, 'such desire may not be motivated by [the masses'] will to overcome alienation, but rather from their more troubling inability to differentiate individual self from social reality (or subjective from objective world)' (Wise 1993/4: 209–10). Following on from this, one can argue that in celebrating the loss of distance, Benjamin has also eliminated the role of the critical subject. Though there may be advantages in technology's capacity to shrink distance and bring things closer, Benjamin's one sided affirmation of this trend meant that:

> the subject was incapable of sufficient distance from the object to experience it dialectically, that is critically as a non-dialectical other, and identity itself became synonymous with the impotence of the subject and the domination of the social system.
>
> (Buck-Morss 1992: 171)

Benjamin's advocacy of the destruction of aura, and its ties to tradition, leaves him with an inadequate theory of contemporary experience. Wise argues that

Benjamin's desire to rid the masses of auratic experience is motivated by 'a Utopian (though misplaced) desire to level *all* class distinctions into one big ontological experience of Being – which would also seem to be coterminous with the proleterianisation of the human race' (Wise 1993/4: 209).

A further critique of Benjamin's faith in this new mode of experience comes from Adorno, who argued against the celebration of a detached, 'distracted' proletariat. For Adorno, the concept of 'objectified subjectivity' is insufficiently dialectical, and fails to account for the way in which proletarian subjectivity is already the product of bourgeois culture. Hence the advocacy of the destruction of aura leads to further difficulties: was starting again with the bare minimum of experience really connected to a rejuvenating assimilation of the past? McCole spells out the dilemma: why prefer the nihilism of Benjamin's new barbarism to that of Jünger's warrior figures if politics comes down to a contest over which of them most decisively liquidates bourgeois culture (McCole 1993: 166)?

The technological destruction of traditional auratic experience does not mean that it will disappear, despite Benjamin's claim for the objective, non-instrumental telos of technology. Indeed, the factors bound up with the concept of aura were translated into myth by the Nazis at the very time Benjamin was writing this essay. The notion of collectivity was distorted for fascist purposes. On this point, Hillach notes the special affinity between film as a medium of reproduction and political mass movements:

> mass movements reflected in this way also became accessible to mass reception for the first time . . . [this quality] made the film suitable for the masses to experience themselves and to enjoy their own mass movements aesthetically. With cameras mounted in front of them . . . the masses can become intoxicated with their subjugation rituals . . . Film reproduction thus becomes a stimulus to political acts that are staged for the sake of their own aesthetics
>
> (Hillach 1979: 118)

Here we have the 'unnatural' use of technology, a rendering of politics aesthetically, and a form of collective reception very different from that which Benjamin intended. The aestheticisation of mass movement politics creates an impediment to political action for, as Hillach remarks 'aesthetic self gratification leaves the masses disempowered of their potential power to change the system' (Hillach 1979: 119). The 'deepening of apperception' in this case has not led to an increased critical consciousness, but rather has been reduced to an aesthetic self-gratification harnessed for fascist purposes.

Such criticisms, however valid, do not sufficiently reveal the political context of the artwork essay, or examine other aspects of Benjamin's thought which develop a more complex position around the emergence of new technologies. Benjamin was aware of the possible betrayal of what he saw as the radical possibilities implicitly inherent in technology. In addition, there were times when he seemed unambiguously to embrace the reduction of older modes of experience that were connected with the oppression of tradition. This aspect of Benjamin's thought needs to be contextualised in the light of the imminence of fascism. If the past was intertwined with barbarism, and the present was undermining the capacity for experience through technology, the only choice for Benjamin seemed be a mode of 'barbarism' that could respond to fascism on its own ground.

Contextualising Benjamin's theory through Jünger

Many aspects of Benjamin's theory of technology, as we have examined it so far, can be regarded as the outcome of his struggle to oppose the right-wing aestheticisation of technology, in particular its more destructive aspects, articulated in the late 1920s and early 1930s. Unlike the cultural conservatives who had railed against technological development, Jünger embraced the speed of technological change as the means to destroy the past utterly and to bring about a whole new mode of existence. Jünger understands technology to be the ontology of modern humanity. It becomes our 'nature'. He predicted the emergence of a 'new, fully technological age, with its joyously anticipated synthesis of flesh and steel, brain and camera, body and machine' (Huyssen 1993: 9).

The aestheticisation of destruction occurred partly through the glorification of war, where Jünger claimed was the ultimate fusion of technology and existential danger. What is particularly evident in his writings of this period is the way his metaphors technologise nature, for example 'storms of steel', 'hurricanes of fire', 'planes as great birds of prey' (Huyssen 1993: 15). This violent metaphorisation reflects Jünger's general attitude towards nature as an inert material that needs to be mastered. As Blumenburg notes:

> [n]ature has ceased to be a binding force for mankind, it has been reduced to the status of a mere object, to be mastered in theory and in practice. In our century, experience suggests that, disappointingly, neither natural materials nor the bodily construction of man can meet the requirements of technological progress. Organic material reveals a distinct inertia ... overcoming this feature was first formulated by Ernst Jünger in the 'organic construction' of the worker.
> (Blumenburg, cited in Wenneburg 1993: 62)

It was this attitude towards nature that Benjamin was to focus on in his response to Jünger, and which led to Benjamin's own incomplete attempt to establish a successful relation between technology and nature.

As well as aestheticising destruction in warfare, Jünger also celebrated the incursion of technology into modern urban life, primarily through the theme of 'danger'. After the First World War, the experience of life in the city became the new site of danger. Jünger argued that modern experience was characterised by the growing incursion of danger into human life, and that the virtue of this development was to shock the bourgeoisie out of its complacency. In an exaggerated masculinist vein, he wrote that:

> [w]e must penetrate and enter into the power of the metropolis, into the real forces of our time; the machine, the masses, and the worker. For here lies the potential energy from which will arise the nation of tomorrow . . . we will try to put aside the objections of a misguided romanticism which views the machine as in conflict with culture.
> (Jünger in Wenneburg 1993: 47)

Jünger's celebration of danger, destruction, war and pain, were only possible through technology. First, there is the notion of the armoured subject: the fascist type of cyborg who is impervious to pain. It is only through the use of techno-armour as shield that Jünger can glorify war and destruction, and come to terms with his own experience in the war (Huyssen 1993). However, second, and perhaps more importantly, technology plays a vital role in allowing lived experience to become abstracted into aesthetic experience. Jünger's aesthetic of destruction is both an aesthetics *of* technology and an aesthetic constituted *through* technology. In order to demonstrate this, it is necessary to examine his conception of a secondary 'technologised' nature.

Technology, argues Jünger, abstracts the subject's mode of perception and allows the 'objectification of reality'. Jünger claims that technology constitutes a 'second consciousness', which allows one to stand outside oneself, to see oneself as an object. Technology provides the armour that allows one to step outside the realm of sensation and become impervious to pain. Such detachment has often been seen as a means for critical analysis, allowing one to step back from immediacy and to grasp the object more perceptively. Brigette Wenneburg compares Barthes' sense of detachment to Jünger, noting that 'while Barthes sees this reflexive detachment as the necessary prelude to critical catharsis, Jünger understands it as no more than a functional adaptation to the omnipresence of danger in the modern world' (Wenneburg 1993: 53). In a curious way, Jünger combines detachment with a heightened sense of affect brought about by the 'extreme nearness' of danger or death.[2] This ability to 'step outside' a former experiential paradigm in order to more effectively grasp the object, is indicative of the means through which technology facilitates a more abstract mode of engagement with the world. As we have seen, this process, for Heidegger, results in the 'age of the world-picture'. Conversely, Benjamin regards the same process as allowing for a more radical consciousness to develop through being able to objectively perceive the world. For Jünger, this detachment allows for the development of the technologised warrior. It manifests itself most clearly through the reconstitution of vision. Throughout his writing, he continually refers to seeing as a mode of attack, photography as a weapon.

Jünger's work functions like a defence mechanism against the shocks and traumas of modern existence, according to Bohrer. Applying Benjamin's formulation that 'the threat from these energies is one of shocks . . . (the) more readily consciousness registers them the less likely they are to have a traumatic effect' (Benjamin, W. 1968: 163). Bohrer argues that Jünger's writing and photographs fend off shocks by rendering them aesthetic, freezing them at a particular point in time (as in photography) and allowing them to be easily registered by the subject's consciousness. Hence Jünger's work registers the fact of danger, but evades the experience of trauma because he 'freezes [such experiences] at the level of conscious perception before their full emotional impact is felt' (Bohrer in Herf 1984: 99). The mechanical eye of the camera allows for a displacement of experience. Jünger's celebration of danger and death can only occur because these experiences take place through a heavily mediated technological process. Subjective experience, when technologically mediated, occurs within a different constitutive frame. The technology that allows for the enhancement of perception also allows for an appropriate means of defence against the shocks of modernity. For Jünger, this abstraction of experience is the natural outcome of our relationship with technology: '[i]t almost seems as if the human being possessed a striving to create a space in which pain ... can be regarded as an illusion' (Junger, in Buck-Morss 1992: 33). Thus, if technology shields the subject from the immediacy of shocks, it also allows an 'aura' of *blitzkrieg* to manifest itself.

Benjamin's advocacy of technological destruction is thus a response to the pressing political situation he found himself in. Obviously he couldn't take the side of the cultural conservatives who rejected technology *in toto*. But was there a way of advocating the new developments in technology, without siding too closely with Jünger? Jünger saw technology as having a telos that led straight to the victory of the revitalised right wing. Benjamin's general response was to argue that, on the contrary, technology unfolded objectively towards a non-instrumental telos.

Benjamin's immediate response came in the 1930 review (Benjamin, W. 1979), where he identified several problems with Jünger's theory of technology. In particular, Benjamin was critical of the aesthetic rendering of technological change, and of couching technology in mythical terms. He writes:

> in the face of the landscape of total mobilisation, the German feeling for nature has an undreamed of upsurge ... the metaphysical abstraction of war professed by the new nationalists is nothing but an attempt to solve the mystery of an idealistically perceived nature through a direct and mystical use of technology, rather than using and illuminating the secrets of nature via the detour and organisation of human affairs.
>
> (Benjamin, W. 1979: 124)

In this passage, we can identify several areas that concerned Benjamin. First, there is the intertwining of technology with myth. Second, there is the lack of temporality, the directness and violent immediacy of technological destruction. These themes will be examined when we return to the question of aura.

Finally, there is the reference to the 'secrets of nature'. What secrets does nature hold, and how can they be helpful in mediating the relationship between mankind, technology, and nature?

Benjamin's short piece 'To the planetarium' gives an indication of how he envisaged the relationship between technology and nature. He compares mankind's relationship to the cosmos in antiquity and the present. In antiquity, he claims, our experience of the cosmos was both ecstatic and collective, whereas modern experience is private and marked by a degree of detachment. What is lost in the modern is the 'knowledge of what is nearest to us and what is remotest to us, and never of one without the other' (Benjamin, W. 1986: 103). Modern experience is typified, he argues, by the manner in which modern astronomy determines our relationship to the stars by privileging the optical sense. Benjamin's claims are interesting for several reasons. The 'correct' relationship to the cosmos, he writes, is marked by both a sense of distance *and* a sense of closeness. By contrast, the technological destruction of aura abolishes a sense of distance, just as does the desire 'of the masses to bring things close', a desire Benjamin appears to celebrate in the 1936 essay. These different approaches can be discussed in terms of the constitutive abstraction argument. To be both distant and close, would mean that the more concrete forms of being (exemplified by the 'ancients') bound the more abstract modes of engagement made possible through modern technology. Benjamin notes the loss that occurs when the optical sense is privileged over a more rounded experience, something that needs to be kept in mind when we examine the privileged technologies of film and photography. Yet if prior modes of being and acting were maintained so that they could resonate sufficiently within practices enabled through modern technologies of extension, this would not necessarily lead to the one-sided privileging of the optical, and thus to Heidegger's 'world picture'. As we shall see, Benjamin's discussion of auratic experience hints at the importance of maintaining these prior settings.

Benjamin warns that we cannot simply disregard as obsolete our relationship to the cosmos, since it is liable to return in the form of technological destruction on a massive scale. This is what occurs with the outbreak of war, he argues, which vainly 'attempt[s] a new and unprecedented wedding with the cosmos'. Similarly those who aestheticise such destruction, such as Jünger and Marinetti, make a 'perverted attempt to re-enact the ancient ecstasies of cosmic experience' (McCole 1993: 187). Benjamin considers these dangerous and unnatural uses of technology. He cites two explanations for this misuse of technology: first, the distorted use of technology under capitalism which only serves 'the lust for profit of the ruling class'; secondly, technology itself, which helps to perpetuate the perverted relationship between technology and nature. He claims that 'the ancient practice of giving oneself over to cosmic experience has been supplanted by an attempt to dominate nature' (McCole 1993: 187). The result is that 'technology has betrayed humankind and transformed the bridal bed into a bloodbath'. The correct use of technology, by contrast, will not attempt to master nature, but to master the *relationship* between mankind and nature.

What, then, is the more 'authentic' experience of the cosmos, that would enable a relationship between mankind, technology and nature not based on domination? The answer may be bound up with the question of aura. Benjamin defines aura, at one point, by reference to nature:

> [t]he concept of aura which was proposed . . . with reference to historical objects may usefully be illustrated with reference to the aura of natural ones. We define the aura of the latter as the unique phenomenon of a distance, however close it may be. If, while resting on a summer afternoon, you follow with your eyes a mountain range on the horizon or a branch which casts its shadow over you, you experience the aura of those mountains, of that branch.
>
> (Benjamin, W. 1968: 222–223)

In many ways, Benjamin's conception of the auratic experience of nature implies its 'humanisation': in one of the essays on Baudelaire he describes the aura as 'a projection of a social experience among human beings onto nature' (Benjamin, W. 1968: 118). However, this humanisation does not seem to arise from an anthropocentric understanding. Benjamin rejected any such assumption along with Adorno's suggestion that humanising nature would involve seeing the 'forgotten human residue in things'. There seems to be a more primordial sense of aura to which Benjamin is referring, one where, according to Rebecca Comay, 'the (re)experience of the aura . . . exceeds the egocentric grasp of a humanistic self-consciousness and indeed points to a humanity beyond self-production and control' (Comay, 1992: 148). Thus, there is a twofold means of distorting auratic experience in Benjamin's descriptions. First, there is the distortion of experience that occurs under the capitalist mode of production. This, however, ties into a deeper sense of distortion, a loss of access to the ecstasies of the ancients and, I shall argue, a severance from a more authentic form of aura. How to consider the *relationship* between humankind and nature is another question, which can perhaps only be approached through the question of social forms and how they structure the experience of aura. This will be considered below.

Aura and its contradictions

Unpacking the concept of aura provides a useful means to define what is at stake in Benjamin's conception of technology, in particular its potential for the positive destruction of the more disempowering forms of aesthetic experience. If we read carefully, we notice an ambivalence towards the destruction of aura even in the 1936 essay.[3] It is worth noting that the purely negative qualities Benjamin attaches to aura in the 'The work of art' essay by no means exclusively defines auratic experience. For instance, Benjamin argues in this essay that aura in aesthetic objects can only lead to a passive experience of the object. Yet in other essays written at much the same time, such as 'The storyteller' and 'Some motifs in Baudelaire', we see Benjamin finding a positive role for communal interaction in contexts that can only be understood as auratic. In addition, as Andrew Arato notes:

> a few works of modern art (those of Kafka, Proust and Baudelaire) . . . manage to combine the very experience of the loss of aura into 'works' suggesting a fabric of residual communication and the anticipation of a qualitatively different society, or at least the reduction of the present one into ruins.
> (Arato 1983: 10–11)

This last point opens up the question of the role of contemporary art. Adorno was critical of Benjamin's one-sided presentation of aesthetic destruction in the 1936 essay, but we can see how in other works Benjamin's position is closer to Adorno's own. Given Benjamin's hostility to Jünger and the Futurists it is perhaps easier to understand why the 1936 essay takes such a position, but as Arato points out, technological reproducibility was not the only manner in which the notions of symbol, false harmony and illusion could be sundered. Benjamin's earlier work showed that allegory also provided an effective means to 'renounce the idea of symbol and beauty'. As such these are also resistant to fascist appropriation and form 'possible contexts of critical and active reception' (Arato 1983: 11).

Benjamin recognised the possibility that technological forms, in relation with the workings of capital, could create new forms of aura. Needless to say, this possibility has been borne out. Yet this ought not cause us to dismiss Benjamin's hopes as a failed historical gamble. Neither do we have to take the opposite course and overemphasise the possibility of resistance embodied in forms of popular culture. Instead, the historical development of new forms of aura invites us to unpack Benjamin's notion even further. In his more positive conception of auratic experience, Benjamin connected such experience to historical traditions and forms of communicable experience. Clearly, the aura that surrounds modern technology, such as film, does not allow this process to take place. As Arato reminds us, 'fake aura does not restore weakened traditions' (Arato 1983: 11). Hence, there must be more to the notion of aura than its existence as a form of authority and mystification.

The historical decline of experience

The part of Benjamin's work we have examined so far tends to advocate an abandonment of older forms of experience deemed auratic. This is because Benjamin sees them as inevitably bound up with the system of domination they evolved from, and helped to sustain. Much of Benjamin's work, however, suggests an alternative project, namely that of reclaiming forms of experience threatened by the dual movement of technology and capital. At times, Benjamin deeply laments the destruction of aura. He even connects aura with the possibility of redemption, writing that 'the decay of the aura and the atrophy of the vision of a better nature . . . are one and the same'. This apparent contradiction can be analysed within the terms of the constitutive abstraction argument. An analysis of social life from the point of view of constitutive forms necessarily implies a distinction between such forms and their specific contents. To argue, for instance, for the importance of the face-to-face as a binding level of social integration is not to, in Sharp's terms 'embrace them as unambiguous ideals' (Sharp 1985: 78). In terms of their specific content we can find violence, as well as love or sympathy. However, I am arguing for the continued relevance of these forms in so far as they provide a setting which contains deeply embedded social and cultural meanings. In relation to Benjamin, I suggest that when he rejects aura as being implicated in a historical system of violence and domination he does so at the level of *content* only. However, we need to examine this 'other side' of Benjamin's work more fully before such a claim can be further elaborated.

Axel Honneth writes that much of Benjamin's work 'attempts to determine a complex concept of experience not in terms of communicative knowledge, but in modes of magical world disclosure' (Honneth 1992: 84). Following from Bergson, Benjamin attempts to contrast this type of disclosure with the purposive-conscious disclosure that governs the majority of activity in the modern world. Tied up with this notion of purposive instrumentality is the experience of shock. In his work on the Paris Arcades Benjamin claims that shock is the essence of modern experience. Buck-Morss argues that 'the technologically altered environment exposes the human senses to physical shocks that have their correspondence in psychic shock' (Buck-Morss 1992: 16). In a manner very different from the 'Work of art' essay, Benjamin writes of the impoverishment of experience in the modern world. Extending a Freudian notion, Benjamin examines how the unconscious defences of the organism protect the subject from shocks emanating from the environment. He notes the manner in which these defence mechanisms are no longer simply personal, but are also mechanical. Mechanical instrumentalisation reconstitutes our mode of engagement with the environment; on the one hand it shields us from shock, on the other it prevents the assimilation of sensations into personal experience.

Benjamin contrasts two modes of experience: *Erfahrung*, where perceptions and sensations become integrated into the subject's experiential field; and *Erlebnis*, where the subject's consciousness responds to shock by blocking out experiences. In the latter case experiences are not retained, and this, in turn, impoverishes the functioning of memory, since 'perception becomes experience only when it connects with sense-memories of the past'. Consequently, the world becomes something merely lived through. Increasingly, humankind can only engage with the world at a superficial level, that of *Erlebnis*, and is increasingly unable to assimilate the data of the world through personal experience.

The increasing instrumentalisation of the environment is bound up with the development of new technology, particularly within capitalist modes of production. Benjamin comments on the change from craft to assembly line production, claiming that this form of '[t]echnology has subjected the human sensorium to a complex kind of training' (Benjamin, W. 1968: 175). Like Heidegger, who argues that the objectification of nature also leads to the formulation of the human subject as standing-reserve, Benjamin claims that industrial technology restricts the subject's ability to derive meaning from their everyday experiences. He comments thus on the subjective experience of the worker on the production line:

> Independently of the workers volition, the article being worked on comes within his range of action and moves away from him just as arbitrarily . . . in working with machines, workers learn to co-ordinate 'their own movements with the uniformly constant movements of an automaton'.
>
> (Benjamin, W. 1968: 175)

Hence the labour of the worker is subject to a kind of experiential alienation, isolation, uniformity, and repetition, in addition to economic alienation. Benjamin noted the manner in which technological devices tended to displace subjective experience, at the same time as they freed them from labour:

> Comfort isolates. And at the same time it shifts its possessor deeper into the power of physical mechanisms. With the invention of matches . . . there begins a whole series of novelties which have in common the replacement of a complicated set of operations with a single stroke of the hand . . . [a]nd besides tactile experiences of this kind we find optical ones as well, such as the classified ads in a newspaper, or even the traffic in a big city.
>
> (Benjamin, W. 1968: 174–175)

Benjamin isolates the double-edged effect of technology: on the one hand empowering the subject via a process of sensorial abstraction, on the other leading to a crisis of meaning as the subject is deprived of significant modes of engagement with the environment. The question remains as to whether Benjamin provided the necessary framework to incorporate this double-edged process into a successful dialectical mediation.

In addition to the phenomenal estrangement in the experience of the production line and the modern city, Benjamin also notes a decline in the experiential dimension of communication. This occurs, significantly enough, through the development of a particular type of reproduction technology, namely the newspaper. In writing of the increasing inability to experience the world subjectively in any dimension other than instrumentally, Benjamin observes that:

> [n]ewspapers constitute one of the many evidences of such an inability. If it were the intention of the press to have the reader assimilate the information it supplies as part of his own experience, it would not achieve its purpose. But its intention is just the opposite . . . to isolate what happens from the realm in which it could affect the experience of the reader.
>
> (Benjamin, W. 1968: 158)

Hence for Benjamin reproduction technology allows for 'the replacement of the older narration by information, of information by sensation [all of which] reflect the increasing atrophy of experience' (Benjamin, W. 1968: 159). Where narratives involved the subject and their experience in the telling of the tale, newspapers tell stories without the need for the concrete presence of the teller or the singularity of their experience. Communication adopts an instrumental character as '[f]ew readers can boast of any information which another reader may require of him' (Benjamin, W. 1968: 159). The range of possible experiences increases in the sense that a reader can learn of an increasing number of events, but such experience also becomes *thinner* when the tangible presence of the other is no longer required in this form of communication.

Technology, in its reproductive capacity and in its ability to bring everything close also rapidly increased the functioning of voluntary memory, which, following Freud, destroyed the fabric of connections linking *Erlebnis* to the kind of richer and deeper experiences that define *Erfahrung*. Thus, in bringing everything closer, technology *also* transforms the bond of closeness in the sense that it simultaneously brings close and allows a certain standing apart. Benjamin's discussion of the technological transformation of experiential categories serves as a particular example of an ontological contradiction. The increasing *range* of possible experiences may lead to the emptying out of the ontological grounds which would allow for any experience to be meaningful. (This possibility will be discussed in more detail in Chapter 7.)

Shierry Weber points out that the result of shock is not only to cut one off from a complex mode of experience, but also from the experience of history, and with it the desire for transcendence.[4] Weber writes that '[i]solated from desire, which depends on the ability to elaborate, through fantasy, wishes that spring from the present but are to be fulfilled through time, one is cut off from the possibility of redemption' (Weber 1981: 266).

One of Benjamin's responses to the separation of *Erlebnis* from *Erfahrung* is to advocate forms of action that occur outside the sphere of instrumentality. These can be grouped around the idea of a lessening of subjective attention to the environment that surrounds it. When writing on topics such as Proust, the flaneur, surrealism or the experience of hashish, Benjamin's common theme is the advocacy of states of consciousness that allow one to rid oneself of instrumental attention. This decentring of the subject is also a 're-animation of reality', a form of magical disclosure where the subject's experience of the world is enriched. The subject/object dichotomy, drawn into even sharper relief by the rise of instrumental modes of being, is, according to Benjamin, dissolved by this loosening of attention. In this state everything becomes 'drawn into the realm of the intersubjective', as the subject experiences the world as 'a social field of analogies and correspondences'. I want to argue that, for Benjamin, such experience necessarily involved the reclamation of a genuine form of auratic experience and furthermore, that without a sufficient account of the forms which carry such experiences, Benjamin faced the danger of lapsing into a kind of mysticism. However, before this issue is taken up, we need to look at a further manifestation of technology, namely the compensatory form of the phantasmagoria.

Experiencing the phantasmagoria

If the experience of mechanised labour and the shocks of modern urban existence cause life to be lived on a fragmented, superficial level, then the phantasmagoria renders such surface experience into a state of illusory plenitude. While technological change produces an experience of shock, it also comes to provide a compensatory mechanism, that of the phantasmagoria, which functions as the dialectical other of the factory. Benjamin defines the phantasmagoria as an appearance, an illusion, which deceives the senses through their technical manipulation. He notes how the increase of technology increases the potential for phantasmagoric effects. In the Arcades Project, Benjamin describes the spread of phantasmagoric forms into public space, where totally simulated environments, such as shopping malls, come into existence. In addition the role of traditional art as a 'sensual experience that differs from reality becomes difficult to sustain' (Buck-Morss 1992: 23).

What is the role of aura in the phantasmagoria? As Buck-Morss notes, it is the 'surface unity which provides the phantasmogoric effect' (Buck-Morss 1992: 33). In this sense the manner in which the auratic work of art (regarded as a unified work) claims its uniqueness and naturalises the social relations of its production, by covering over the actual fragmented state of social relations, resembles the way the illusory wholeness of the phantasmagoria covers over the real state of both subjective and social fragmentation. Some aspects of the phantasmagoria resemble the function of the work of art. If technology allowed for the creation of a 'phantasmagoria of textures, tones and sensual pleasure that immersed the home dweller in a total environment, a privatised fantasy world that functioned as protective shield for the senses',[5] then this resembles Benjamin's description of the auratic work of art in a number of ways, in particular both allow a process of beautiful illusion, absorption and mystification.

In the Arcades Project, Benjamin is much more ambivalent about the role of technology. He recognises its double function: on the one hand, technology extends the human senses, increases the range and depth of perception and 'forces the universe to open itself up to penetration by the human sensory apparatus'; on the other, the technological extension of the senses leaves them open and vulnerable, in the sense that technology 'doubles back in the form of *illusion*, taking over the role of the ego in order to provide defensive insulation'

(Buck-Morss 1992: 33). As we shall see, Benjamin hoped that film and photography might be able to break apart this illusion, by using the very technology that perpetuates shock and the compensatory form of phantasmagoria against itself.

Benjamin's project is, then, according to Buck-Morss:

> to undo the alienation of the corporeal sensorium, to restore the instinctual power of the human bodily senses for the sake of humanity's self-preservation . . . not by avoiding the new technologies, but by passing through them.
>
> (Buck-Morss 1992: 5)

Here lies the potential of this aspect of technology. I want to examine this potential by developing the theme of the 'optical unconscious'.

The optical unconscious of film and photography – mimetic technique

While technology had radical possibilities in its capacity to defamiliarise habits of perception through the dislocation of the senses, Benjamin was acutely aware of how the discursive identity of the modern subject could all too easily be harnessed, more than ever, towards the logic of capital. The abstraction of sensation left the subject without a ground for resistance, and engagement with the world increasingly took place through instrumental means, as new forms of power arose to manage the abstracted and formalised technological senses. Jonathan Crary succinctly sums up the dilemma of the technically mediated subject in relation to the abstraction of the visual sense:

> But almost simultaneous with this final dissolution of a transcendent foundation for a vision emerges a plurality of means to recode the activity of the eye, to regiment it, to heighten its productivity and to prevent its distraction. Thus the imperatives of capitalist modernisation, while demolishing the field of classical vision, generated techniques for imposing visual *attentiveness*, rationalising sensation and managing perception.
>
> (Crary 1992: 24)

Crary stresses that these new instrumental techniques required 'a notion of visual experience as instrumental, modifiable and essentially abstract (which) never allowed a real world to acquire solidity or permanence' (Crary 1992). Although Crary only discusses vision, one could easily extend his comments to all technologically mediated sensations. The reconstitution of the visual sense serves as a particular example of how the dominance of a more constitutively abstract mode of practice can lead to the instrumentalisation of social life.

Benjamin found, however, that the new technologies could also provide a means of resisting these instrumental tendencies. He claimed, for instance, that photography could grant access to involuntary memory, which connected surface impressions to *Erfahrung*. In a 'Small History of Photography', Benjamin suggested that 'the camera [is] . . . ever readier to capture fleeting and secret moments whose images paralyse the associative mechanisms in the beholder' (Benjamin, W. 1986: 256). Such a process would allow the flood of sensations that emanate from the modern world to be stopped at some point, to crystallise in a monad-like impression that could be properly assimilated to experience. How would such a process occur? One answer lay in developing the role of the 'optical unconscious'.

The notion of the 'optical unconscious' sits uneasily beside the other aspects Benjamin lists as radical possibilities inherent in the filmic medium. For instance, Benjamin claims that film can begin to examine reality with a 'scientific, politically critical eye'. This claim appears to be how Benjamin distinguished the cinema from the arcades. He did not regard the cinema as a site of the phantasmagoric, but rather, as having the potential to break apart the illusion of everyday life, 'bursting the prison world asunder'. However, Benjamin did not simply advocate film for its ability to penetrate more deeply into reality. One need only recall that, in his essays on Baudelaire, he insisted that 'the kind of memory that can lend coherence to ongoing, accumulating experience depends on associative richness rather than mechanical precision in recalling discrete facts' (McCole 1993: 7).

Overemphasising the importance of cinema's attention to scientific detail would only further the encroachment of instrumental forms of attention, a phenomenon Benjamin correctly identified as one of the sources of subjective disempowerment. On this point, Virilio remarks that: 'too much exactitude results in the definition of the recorded and transmitted form-image – results in inexactitude, or better, a relative uncertainty due to the interpretative delirium of the observer' (Virilio 1991: 75). While Benjamin praised technological devices for their capacity to examine reality more comprehensively, this in itself was not sufficient to generate the radicality of the filmic medium. On the contrary, this process *simply by itself* might only lead to an increase in instrumental consciousness, or *Erlebnis*, as the subject became bombarded with more reality, more shock. While film generated the ability to consider the world from a more abstract vantage point, the generation of 'associative richness' could only occur through reference to forms of experience that lay outside this vantage point. In other words, the relevance of prior constitutive forms lies in their being able to provide a semantic resource that more abstractly constituted practices can draw upon.

Though Benjamin at times praises film for its ability to give us greater mastery over our environment, the significance of this needs to be considered carefully. One need only reflect on Benjamin's use of the term 'optical unconscious'. The radicality of the unconscious lies in its capacity to undo forms of mastery. As such, the scientific qualities of film and photography occupy a curious place in Benjamin's schema. However, if we reflect that today the discourses of objectivity and scientific exactitude have been revealed to constitute an aura of their own, then we can further circumscribe this element in Benjamin's work, and argue that there is much more to seeing through aura than a testing, objective attitude.

What then, distinguishes the cinema from the arcades? The answer must lie in the fact that the mode of perception enabled through cinema can be accompanied by a form of distracted, non-instrumental attention. Here, distraction would involve a forgetting, but a very different mode of forgetting than is involved in the fetishising of the auratic work of art. The object of this mode of forgetting is the transcendence of instrumentality, where instrumentality involves the constitution of social life on a single plane. This is how technology operates through the phantasmagoria. Distraction provides a means to move from the phantasmagoric to a different mode of engagement.

Benjamin hoped that film might be able to tap into an *ontologically prior* form of auratic experience, the form Benjamin gestures toward whenever he discusses the relationship between technology and nature. In the case of film, this form of aura is bound up with Benjamin's notion of mimesis. Benjamin's project would involve the mimetic capacity of film to use the violence of the modern shock experience 'against itself'. Benjamin regarded film as a 'mimetic technology' which 'provided an expressive medium adequate to industrially transformed perception'. The potential of film lay in its capacity to recapture a sense of auratic perception, however one that did not lapse into authority, mystification, or fetishism. This redemptive project involved triggering off forms of collective memory and experience unavailable to forms of instrumental perception. As Hansen points out, 'the mimetic capabilities of films . . . [would] not only fulfil a critical function but also a redemptive one, registering sediments of experience that are no longer or not yet claimed by social and economic rationality, making them readable as emblems of a forgotten future' (Hansen 1987: 209). However, as Hansen notes, to recapture the returning gaze through the filmic medium would only be possible if film provided a suitably unsettling, or transgressive experience, through the use of montage and discontinuity rather than mainstream plot and editing techniques. Only then might the mimetic form of shock be able to tap into the layers of 'unconscious memory buried in the reified structures of subjectivity'. The potential of cinema to uncover the buried collective memory of its subjects thus underlies the sense of collective reception to which Benjamin refers.

Mimetic shock could also help to undo the reification of time. We have seen how much of the experience of modernity was marked by an eternal recurrence of the same, the experience of modern forms of work, for instance. In the introduction I discussed the way the subjective experience of time is constituted at varying degrees of abstraction. Within modernity, we can observe the increasing dominance of a more abstract framework through which temporal experience is constituted. At this level, subjective practices are primarily valued in relation to how quickly they can achieve a set goal or end. Time then, becomes an external constraint to be overcome. This approach to time is most fully realised in the Futurist project, in particular, their worship of pure 'speed'.

Such temporal reification, according to Hansen, erodes 'the capability and communicability of experience – experience as memory – as awareness of temporality and mortality – but the very possibility of remembering imagining, a different world' (Hansen 1987: 189). The use of cinematic technique, montage, discontinuity, fragmentation, might help mimetically to initiate a differentiated experience of time to the subject. This leads us to a vital question, namely, can the problems Benjamin saw in the historical loss of experience, or in the misuse of technology, be solved simply by correct *technique?*

As we have seen, the cinema destroys the aura as a form of false trace, but also has the potential to produce a different, more positive form of aura through the creation of the 'optical unconscious'. However, I want to argue that this reproduction of aura can all too easily be fetishised, and degenerate back into the form of false trace if its insights are *not reflexively read back into forms of being that occur outside of the technological sphere.* The work of Crary has shown us how the abstraction of the senses can lead to increasingly instrumental modes of being. He notes that the technologically rendered visual world acts to relocate vision to a place outside the observer. As such, the *meaning* of vision is radically different (Crary 1992: 1). The disruptive potentials suggested by film may allow for some form of alterity, but such an experience always runs the risk of being transformed into a fetish, if it cannot be structured by a social form that could translate the 'experience' of aura into a more concrete and stabilised mode of engagement. This issue will be further considered in the conclusion.

Conclusion – the social structure of auratic experience

Can one hold together the negative and positive readings of aura so that humanity does not merely shake off experience and tradition, but engages with them more productively? Gill and Riggs have noted that Benjamin's 'notion of aura as elitist seems to parallel Heidegger's idea of inauthenticity: presence is inauthentic, yet the process of presencing or unconcealedness *is* authentic' (Gill and Riggs 1995: 93). We can draw from this point the vital distinction between authentic forms of being and their inauthentic traces, or contents. This allows us to join together the disparate parts of Benjamin's writings on technology and experience, and to develop a criterion from which an ethical relation to technology might begin to emerge.

We can begin to attempt such a relation by looking at Benjamin's comments on the aura and the reciprocity of the gaze. In the essay 'On Some Motifs in Baudelaire', he writes:

> looking at someone carries the implicit expectation that our look will be returned by the object of the gaze. Where this expectation is met (which

in the case of the thought processes can apply equally to the look of the eye or the mind and to a glance pure and simple), *there is an experience of the aura to the fullest extent*. 'Perceptibility' as Novalis put it, 'is a kind of attentiveness'. The perceptibility he has in mind is one other than that of the aura. . . . The person we look at, or who feels he is being looked at, looks at us in return. To perceive the aura of an object we look at means to invest it with the ability to look at us in return. This experience corresponds to the data of the *mémoire involontaire*.

(Benjamin, W. 1968: 188)

Several things can be developed from this passage. First we can note how different this description of auratic experience is from that discussed in the 'Work of art' essay. Here, we have a form of aura capable of invoking a certain type of mood, a kind of attentiveness. Instead of absorption, fetishisation and domination, we are told that aura allows for a kind of reciprocity, some form of intersubjective relation. Second, Benjamin tells us that this 'is an experience of aura to the fullest extent'. Benjamin appears to be formulating a definition of a more authentic form of aura that is bound up with the issue of reciprocity, of 'returning the gaze'. Of course there are varying modes of reciprocity. Is this a form of reciprocity that would reduce the other to the same (such as Heidegger's *Gestell*) or a more general economy that would resist a form of subjective appropriation and closure? Finally, what is the significance of the tangible presence of the other for the capacity to experience the aura to its fullest extent?

Arguably, Benjamin meant for aura to have the possibilities of the latter mode of reciprocity, functioning to disrupt the economy of the same. In his essay on photography, Benjamin compares the appropriative form of reciprocity that occurs in commercial pictures, the '"they're looking at you" of animals, people and babies, that so distastefully implicates the buyer' (Benjamin, W. 1986: 251), with the mode of reciprocity that occurred in the early days of portrait photography. Benjamin quotes Dauthendey, one of the first of these photographers:

we didn't trust ourselves . . . to look long at the first pictures [we] developed. We were abashed by the distinctiveness of these human images, and believed that the little tiny faces in the picture could see *us*.

(Dauthendey in Benjamin, W. 1986: 251)

In this case we have a non-appropriative form of reciprocity, where the subject cannot be consumed in the look. Like Heidegger, Benjamin contrasts a form of dominative looking with a more ontological kind of vision, a fleeting, non-totalising 'glancing'.[6] What is important to note is how this mode of reciprocity manifests itself. The lack of appropriation, the respect for the other borne out in Dauthendey's statement draws upon a form of interaction based in *presence*. Looking as if the subject was really there, one cannot consume the other so readily. Dauthendey's impression serves as an example of the continuing relevance of the face-to-face integrative form, where the generalised sense of the other person is only abstracted with some unease about its consequences. We can link this sense of unease to the broader question of reflexivity, and how we negotiate the relationship between different constitutive forms.

In *The Origin of German Tragic Drama*, Benjamin writes: 'that which is original . . . is never revealed in the naked and manifest existence of the factual' (1977: 45). We can extend this to argue that aura functioning to the 'fullest extent' *relies upon* a constitutive framework that allows particular modalities of experience to unfold. Andrew Benjamin argues that the experience of the aura is the experience of an expectation, or possibility of the potential that lies in the face of the other. This possibility 'introduces the possibility of an ethical dimension . . . into what had hitherto been straightforward questions of aesthetics' (Benjamin, A. 1991: 147). The focus on the intersubjective realm of the aura opens out into the question of semantic survival. It is the auratic experience that preserves meaning by projecting it towards the face of an(other). In other words, the strongest variety of aura occurs when the face-to-face level of social integration binds relations carried out within more abstract levels.

This is the point Benjamin makes in 'The Storyteller', when he distinguishes between the story and information. The former is endowed with the ability to survive, while the latter often does not survive past the moment of its consumption. Like the mode of looking that respects the other as if they were really there, semantic survival draws upon social forms that are based in the tangibility of intersubjective relationships. When aura lapses into its degraded form, this is arguably because these prior forms of the social are transcended: looking becomes voyeuristic appropriation, stories become information.

This reading allows us to understand the possibilities inherent in auratic experience. It prevents us from taking a one-sided reading of Benjamin on technological destruction. We can read Benjamin's 'other side', his stress on the importance of the aura 'beyond the melancholy interplay of nostalgia and redemption' (Benjamin, A. 1991: 150). Such a reading may begin to explain the contradiction whereby the death of aura is welcomed as an integral part of Benjamin's utopian project, while at other times aura is revived for the same utopian ends in terms of a distinction between social forms and their specific contents. The question that remains for us, the question Benjamin never solved, is how to develop a theory of technology that retains the ethical intersubjective qualities of the aura, without its repressive aspects.

Discussing the notion of a more authentic, primordial notion of aura, one continually faces the danger of falling into mysticism. As Habermas has shown, Benjamin never really reconciled the polarities of mysticism and enlightenment (Habermas 1983). Unpacking Benjamin's notion of aura ensures that an essential place must be reserved for conserving forms of experience deemed auratic. I have argued that 'experience' is only meaningful in relation to the constitutive forms through which it is carried. The dilemma of technology lies in the tendency to subsume all forms of experience into its sphere, the result of which actually displaces the normative grounds through which auratic experience could be significant. For Benjamin, we can see such a process in the phantasmagoria, in Jünger and Marinetti's aestheticisation of destruction, and in the increase of shock. What is common to these is the ontologisation of surface, a thinned out and abstracted modality of worldly engagement. Where Benjamin finds a radical potential in technology, it is possible to say two things. First, when Benjamin celebrates the destructive capacity of technology, he is, as we have seen, making a distinction between authentic traces and false substitutes. Arguably this is also a distinction between social forms and their specific contents. While technology allows for the revealing and overcoming of such fetishised content, this ought not to come at the expense of the social form that carries it. Otherwise, stories become information, mimetic fragmentation becomes 'shock'. Hence, the second point: if technology allows for authentic modes of auratic experience to unfold through various techniques, then it can do so only by existing in a reflexive relationship between the relatively concrete forms that allowed the original modality of aura to be revealed and the more abstract forms constituted through technology. Only then might the experience of aura be structured in such a way as to prevent its lapse into false traces.

4 Futurism and the politics of a technological being in the world

It is clear that the Italian Futurist movement was concerned with and shaped by the developing technology of the time. From the manner in which media technology assisted the creation of Futurism as a cultural movement to the love of speed, the machine, and the technology of warfare, Futurism directly engaged and embraced the increasing technologisation of the world, both in theory and in practice. Marinetti was an expert manipulator of the media. By placing the first manifesto of Futurism on the front page of *Le Figaro,* Europe's premier newspaper in 1909, he cleverly used the technology of the time to create a new cultural movement that reached a mass audience virtually instantly. Futurism was as skilful in using new technology as it was in theorising its possibilities or worshipping its beauty. However, I want to argue that the Futurist relationship to technology is much broader than the simple valorisation of this or that weapon or machine. The technological aspect of Futurism went beyond mere 'automobilecentrism', as Wyndham Lewis derisively characterised it.[1] The love of the machine is representative of a more general desire, on the part of the Futurists, to transcend the socio-cultural framework of the period. In this respect, the love of technological speed, the call for the abolition of syntax, the project for the destruction of the 'I' in literature, the repeated desire in Futurist writings to transcend gendered and bourgeois forms of identity, and the call to embrace war, all form part of a generalised sensibility that finds liberation through a more abstracted mode of being in the world. The Futurists envisage a form of social and cultural transcendence enabled by the more abstract reconfiguration of social life that I have argued characterises the dominant framework through which technology impinges upon our worldly activities. To this extent, Heidegger's remark that 'the essence of technology is nothing technological' (Heidegger 1993: 329) is especially apposite in the case of the Futurists. Whether worshiping vehicular speed or valorising the destruction of syntax, the 'essence' of Futurism is framed by a general sensibility that can be characterised as technological.

The founding manifesto of Futurism sets out the basic thematic elements of the movement. While hoping to eradicate all myths of bourgeois life, the Futurists set up a new mythology of their own, formulating a new mode of being, which embraced the potential of new technology to obliterate past and present ways of life deemed passé. Broadly, one can determine several thematic elements in the founding manifesto that define the Futurist project (Marinetti in Flint 1972: 47–52). First, what is particularly clear is the love for technology, especially in its ability to create 'speed'. The cult of speed manifests itself most often through the worship of the automobile. Second, the Futurists emphasise the need to reopen access to certain 'primordial forces', a need which, paradoxically, only technology can meet. Hence, the Futurist automobiles have the quality of ancient 'snorting' and 'torrid' beasts. The industrial sludge from a nearby factory, which Marinetti lies in after his car accident, provides the means through which his primal passions are re-awakened as he thinks of 'the blessed black breast of [his] Sudanese nurse'. Repeatedly the Futurists valorise technology for its ability to transcend what they regard as the languid structures of contemporary society and to provide direct access to these primordial forces. Third, they subscribe to a close identification of struggle and beauty. We are told that '[e]xcept in struggle there is no more beauty' and that '[n]o work without an aggressive character can be a masterpiece' (Flint 1972: 49). Instead of the traditional equation of harmony and beauty, we are to find beauty only in the experience of struggle and disharmony, whether through war or the creation of new forms of culture. Fourth, a general contempt for women pervades Futurist thought. Finally, there is a rejection of the very co-ordinates which have traditionally framed subjective experience. For the Futurists, 'Time and Space Died Yesterday' (ibid.). This overwhelming desire to shed the co-ordinates of time and place sits uneasily with the repeated commitment to the regeneration of Italy as a heroic nation (hence committing themselves to a very specific notion of space). We shall examine this contradiction below.

These seemingly disparate themes are all framed by a sensibility that can be termed technological; that is to say, a reconstitution of previous modes of being within a more abstract framework made possible by technology, but determining a certain mode of engagement with the world irrespective of whether any particular technology is used. Following on from Heidegger we can see how Futurism is a continuation of the historical and metaphysical trend towards an increasingly subject-centred mode of engagement with the world. Aspects of this metaphysic are already apparent in the opening manifesto, with its overwhelming subjectivism, predicated upon a domination of otherness, in particular women and nature. Yet two important things need to be noted here. The first is that *historically*, this period represents the first occasion when the metaphysical underpinnings that shape the Futurist project were able to be realised to any significant degree. Prior to the rise of modern technology, there were 'natural' limits to the operation and consequence of any subjectivist metaphysics. In the early twentieth century, however, new technology allowed an increased overcoming of these limits. The car and the aeroplane, the cinema and the wireless, the new technologies of modern warfare, created the real possibility of 'overcoming', or at least radically reconstituting, our relationship to time and space. This process also enabled an increasing sense of subjective power, since personal experience is less constrained through natural or cultural limits and is extended through technological means.

The second point concerns the operation of this metaphysic. The metaphysical impulse that underlies the Futurist vision is fed through new technology. Technological mediation works to constitutively abstract the subject from previously established modes of being in the world. It is not so much that forms of otherness are eradicated, but rather that they are *reconstituted* though a more abstract framework. The Futurists' commitment to a technological metaphysic also closes off access to prior modes of being, which were aggressively rejected. This chapter examines how this particular type of constitutive abstraction and reconstitution functions within their writing, and how this new 'technological' mode of engagement feeds the ideology of fascist aggression and patriarchal power, which forms the dominant Futurist sensibility. The aggressive closure of previous modes of engagement with the world means, in the context of Italian Futurism, that these repressed modes return to violently re-assert themselves. This is particularly notable in the case of the subject and the nation-state. Having theorised a technological mode of being in the world which would effectively undermine the existence of both the classical subject and the nation-state, these return with a vengeance, their identity violently reasserted through an aggressive domination of the other. Yet while it is clear that Futurism was co-opted within the Italian fascist sensibility, it is also possible to argue that their ideological fantasies, structured through a technological process of socio-cultural abstraction found ultimate consummation within the framework of modern capitalism.

It is here that I would wish to take issue with those who distinguish several movements within Futurism, for the purpose of defining the degree to which they resist or capitulate to Italian fascism. This position argues for an initial period (about nine years) of heroic and relatively autonomous Futurism, followed by a successive period or periods of decline, from the point where a debased Futurism succumbs to the political ideologies of fascism.[2] Such accounts are problematic first, because they elide the thematic continuities between early Futurism and its later forms, and second, because they fail to appreciate the degree to which the framing sensibilities of Futurism, even in its heroic phase, contribute to the fascist project of social and political regeneration within Italy. These issues shall be dealt with in a moment.

The operation of technological metaphysics

Two statements by the Futurists help to convey the changed sensibility resulting from the impact of new technologies, a sensibility enthralled by the radical explosion of new possibilities. The first recognises how technological change necessarily results in broader cultural shifts that determine how we engage with the world. In 1913 Marinetti wrote that:

> Futurism is grounded in the complete renewal of human sensibility brought about by the great discoveries of science. Those people who today make use of the telegraph, the telephone, the phonograph, the train . . . do not yet realise that these various means of communication transportation and information have a decisive influence on their psyches.
>
> (Marinetti, in Appolonio 1973: 105)

More directly, Marinetti wrote of the exciting possibilities of being able to connect across time and space:

> An ordinary man can in a day's time travel by train from a little dead town of empty squares . . . to a great capital city bristling with lights, gestures and street cries. By reading a newspaper the inhabitant of a mountain can tremble each day with anxiety, following insurrection in China, the London and New York suffragettes . . . The timid, sedentary inhabitant of any provincial town can indulge in the intoxication of danger by going to the movies and watching a great hunt in the Congo . . .
>
> (ibid.)

In this statement, it is possible to understand an essential ambivalence surrounding technological change. Clearly, there is much of value in the manner through which technology enables us to transcend the parochial. To be able to go beyond the local and to be aware of others leads to the possibility of extended relations of co-operation and the construction of a globalised ethic. Yet we can see already a different approach emerging in the Futurists, one which frames the experience of extended technological relations entirely from within a self-framing mode of subjectivity. While the Futurists' subject trembles with excitement or becomes intoxicated by distant dangers, the other exists as material to nourish the will to power. Thus Marinetti reveals the Futurist celebration of technology as framed by a subjectivist metaphysic. This is even clearer when, affirming the 'New Religion-Morality of Speed', he offers us this striking image:

> The opaque Danube under its muddy tunic, its attention turned on its inner life full of fat libidinous fecund fish . . . [h]ow long will this shuffling stream allow an automobile, barking like a crazy fox terrier, to pass it at top speed? I hope to see the day when the Danube will run in a straight line at 300 kilometres an hour.
>
> (Marinetti in Flint 1972: 103)

If we need to be convinced of the metaphysic at the heart of the Futurist vision, we might want to recall Heidegger's classic image of the operation of the *Gestell*, that of the Rhine dammed and at our sole command (Heidegger 1977: 297). Marinetti operates entirely within this frame, regarding the Danube as pure standing-reserve. Nature is objectified as resource material for the projects of humankind. Indeed, nature is abstracted as part of the play of culture. The Futurist project of individual and national regeneration was thus predicated upon the domination of the other. So while for Appolonio, 'the futurists saw the practice of art as a font of energy . . . to such an extent that no productive element of the environment should remain untouched by it' (Heidegger 1977: 297), we need to keep in mind that this dynamism is wholly subject centred and aggressively hostile to any form of otherness.

In relation to this, it is no surprise that flying became a privileged activity for the ecstasies of the Futurist experience. Marinetti writes:

> [a]s I looked at objects from a new point of view, no longer head on or from behind, but straight down, foreshortened, that is, I was able to break apart the old shackles of logic and plumb lines of the ancient way of thinking.
>
> (Flint 1972: 88)

This passage provides a key to the link between abstraction and the will to transcendence that governs all Futurist thought. While here the process of looking down from above and taking apart might be literal, it also serves as a metaphor for all Futurist activity. The idea of looking down from above is a central metaphor of abstraction. For the Futurists, subjective power was obtained through looking down from a more abstract vantage point to allow a more totalising vision to unfold. The Futurists embraced technology precisely because it could facilitate this process: the technological overcoming of time and space allowed for the unfolding of conditions that would help further the Futurist will to power.

The cult of 'speed'

Speed occupies a central role in the Futurist project. Speed is the condition wherein a 'new beauty' can take hold in the world and it is valorised by Marinetti as a new 'religion-morality'. Speed defends man from memory, slowness, analysis and habit. It conquers time and space. Speed is pure – it is a powerful process of synthesis which reconstitutes other forces within its framework. It shows 'a scorn for obstacles' and provides a 'desire for the new and unexplored'. To the extent that speed can purge all forms of otherness, it is described by Marinetti as 'hygiene'. Finally, speed allows for an experience of the divine. Marinetti writes that 'the intoxication of great speeds in cars is nothing but the joy of feeling oneself infused with the only *divinity*' (Flint 1972: 102–104). In a metaphor describing the destruction of time and space by pure speed, Marinetti predicts that stable places of human dwelling, houses and cities, will be destroyed, in order to make room for transient meeting places for cars and aeroplanes.

Speed effectively encapsulates the Futurist ecstasy in transcending the moribund structures of contemporary life. Speed overcomes obstacles, namely the prior frameworks that structured social experience. Speed destroys all points of stability and habitation, from the more tangible (houses and cities), to the ontological co-ordinates of time and space itself. Technology facilitates speed: it is what allows the subject to be lifted out from older settings and to inhabit this new plane. The mode of engagement created through speed is regarded as uncontaminated by more worldly things: it is pure and hygienic, devoid of the necessary time and space that could provide reflection, anticipation, memory, even perhaps thought itself. Indeed this is why speed is described in terms of intoxication, religious or otherwise, since the joy comes from removing the subject from anything that could provide space for thought, reflection or even responsibility. These are lower things, sentimental trappings to be cast off as quickly as possible. The 'new beauty' made possible by speed is a harmony arising from the eradication of all forms of otherness, whether external (we are moving too fast to stop and engage with them) or internal (as the space for thought, reflection, anxiety is removed). As Marinetti writes: 'We already live in the absolute because we have created omnipresent speed' (Flint 1972: 49). Speed is valorised as a condition free of ambivalence or contradiction.

Today, those who write of speed are not so certain of its benefits. Sixty years after Marinetti, Virilio writes of technological speed, describing its operation and effect in similar ways. However, Virilio does not see speed as leading to a process of regeneration, but rather condemns its operation as leading to a crisis of meaning brought about through the abolition of ontologically significant habitats (see Chapter 6). By contrast, the Futurists disavowed any crisis of meaning by recourse to violence and domination, thus creating the conditions whereby fascism can readily take the place of the lost ground or habitat, replacing it with a mode of dwelling based on the more violent forms of speed.

One of the prime sites for the operation of technological speed was the emergent cinema. The new technology of the cinema allowed the reconstitution of our experiences of time and space. Hence, cinema was regarded as a prime site for the Futurist revolution, because of its ability to facilitate a 'joyful deformation of the universe'. Again, however, it is not just destruction the Futurists desire, but rather the reconstitution of the world and the power that results from that reconstitution. In 'The Futurist Cinema' (Flint, 1972), the cinema is praised because it allows the 'SIMULTANEITY and INTERPENETRATION of different times and places' (Marinetti in Flint 1972: 141). Stephen Kern notes how the Futurists praised the cinema for its ability to defer the normal forms of temporality, by thickening the present. For the Futurists, cinema gave 'the intelligence a prodigious sense of simultaneity and omnipresence'. The Futurists hated the sequential nature of books. Cinema, by contrast, gave a 'fleeting synthesis of life in the world' (Kern 1983: 172).

The cinema allows the reproduction of images and the simulation of experiences detached from their sites of production. It allows the reconstitution of time, by increasing the degree of experiential material available at any one time (e.g. showing three films simultaneously, speeding up the flow of images, montage effects). Much like the claims made for virtual reality in the 1990s (see Chapter 7), the Futurists valorised the cinema because it allowed more subjective control over the experiential environment. In concluding their manifesto, the Futurists declare that the potential of the cinema lies in how 'we decompose and recompose the universe according to our marvellous whims' (Flint 1972: 142). This declaration describes the central underpinning of the Futurist movement, the ability to break down the traditional frames of worldly reference and reconstitute them according to the will of the subject.

Superficially, Benjamin's work on the cinema might seem to resemble the Futurists in that he also hoped that it might trigger off experiences closed to the contemporary subject. But there is a world of difference between Benjamin's qualified reclamation of aura and the Futurist worship of primal and destructive experience. The Futurists wished to harness the reconstitutive ability of the cinema to the power of the subject's will, to remake the world according to their own desires. Benjamin, by contrast hoped that the cinema would be a redemptive (as well as critical) experience, accessing forms of memory and collective experience outside the parameters of social and economic rationality. Benjamin's theory of the cinema, with its emphasis on collective reception, memory, and critical distance, sharply contrasts with the Futurist position, which frames all experience in terms of the subject, rejects memory as nostalgia and contemplation as cowardice. Additionally, as I have argued, Benjamin's work on the cinema carries an implication that experiential fetishism can only be avoided if the experience of cinema exists in a dialectical relationship with forms of being outside of its sphere. By way of contrast, the Futurists precisely make a fetish of cinematic experience and other experiences of this kind.

In addition to praising the cinema, the Futurists went on to endorse other processes which unhinged the necessity of connection between actions and particular locations. Marinetti claimed a new beauty was made possible through 'the simultaneity that derives from tourism, business and journalism' (Flint 1972: 140). Modern technologies facilitate such processes. New forms of transport and communications breach the limits of time and space. What is interesting about this remark is that the character of simultaneity of tourism, business, and journalism is achieved through a process of constitutive abstraction. Tourism allows the abstraction of subjects from their normal place of dwelling, so that they can reside in new places. The cultural relationship to specific places is abstracted as tourists exchange one place for another. The types of relation enabled through modern business are more abstract than prior forms of exchange, since the commodity exchange enables a more abstract mode of integration than in barter exchange. Journalism too, allows for the experience of things and places outside of their place of origin. The beauty to be found in all three is the experience of flux and exchange, enabled by the abstraction of experiences from places. Technologies like the motor car, the cinema, the radio, all allowed for a transcendence of local, embedded historical frames of reference. Moreover, this sensibility was carried over into other areas of life.

For instance, in 'The Destruction of Syntax – Wireless Imagination – Words In Freedom', Marinetti outlined the bold new aesthetic the Futurists were to hold. In contrast to what he regarded as the fetishisation of single words in Mallarmé's poetry, Marinetti wished to transform literature into pure speed, the rapid flux of rhythms and random words forming a new aesthetic model. He writes:

> I oppose the decorative, precious aesthetic of Mallarmé and his search for the rare word, the one indispensable, elegant, suggestive, exquisitive adjective. I do not want to suggest an idea or sensation with the purest airs of graces. Instead I want to grasp them brutally and hurl them in the reader's face. Moreover, I combat Mallarmé's static ideal with this typographal revolution that allows me to impress on the words (already free, dynamic, and torpedo-like) every velocity of the stars, the clouds, the aeroplanes, trains, waves, explosives, globules of seafoam, molecules and atoms.
>
> (Marinetti in Appolonio 1973: 105)

Two things are to be noted here. The first is how the destruction of syntax, of grammatical rules and conventions, resembles the process of technological mediation. Just as communication and transport technologies allow one to range across the globe and to break it down into a series of optional images or locations in the absence of their prior contexts, so too does the abolition of syntax fragment language into image-like pieces to be reassembled at the subject's will. On this point Peter Nichols comments that 'the scatter-gun effect of Futurist analogies produces a horizontal cross-weave of images which brings together different, initially unrelated zones of experience' (Nichols 1989: 207). The breaking down and reconstituting of unrelated zones of experience was extended beyond the realm of art, to become the central form of experience itself. Hence Marinetti's enthusiasm for new communications technologies that would allow the reorganisation of once embedded modes of experience across a variety of regions, according to the will of the subject.

Second, Marinetti constructs a type of one-dimensional aesthetic theory that relies on superficial effects *at the level of the text*. Whereas Mallarmé's aesthetic works to evoke a 'prior' experience,[3] Marinetti's deliberately anti-contemplative aesthetic evokes nothing except the experience of the self-present Futurist text. He does not wish to use language to engage with a more concretely framed experience, but rather to situate all experience entirely at the more abstract level of the Futurist text. Nichols has identified a gendered ideological quality in this move, seeing the Futurist will to abstraction as a denial of 'feminine' qualities of materiality and inwardness. He comments that 'Futurism's preoccupation with speed and simultaneity derive not simply from an obsession with technology, but from a need to find aesthetic means by which to deny linguistic materiality as the province of a feminine inwardness' (Nichols 1989: 207). There is no doubt that much of Futurism's will to abstraction stemmed from a denial of the feminine, to the extent that they are understood as contaminating the hygiene of pure speed. The category of the feminine itself becomes more abstract within Futurist discourse, transposed onto the worship of feminised machines.

It is worth pointing out that the Futurists' position towards women is at times more ambivalent than simply a denial of the feminine. For instance, they opposed bourgeois traditions such as marriage and protested against the traditional place of women in Italian society. The 'tactile' manifesto, which we shall examine later, proposes a non-phallic polymorphous model of sexuality, which would resonate with some strands of contemporary feminism. It would seem then, that the 'feminine' is opposed to the extent that it stands in the way of the possibilities for transcendence. Marinetti would have undoubtedly concurred with the sentiments of much of 'cyber-feminism', whereby the Internet empowers women through its ability to transcend the material and gendered world. The comments of one 'geek girl' along the lines that 'tech will solve all our problems . . . Girls *need* modems', is indicative of this general equation of technological transcendence with individual empowerment. Moreover, the sentiment that '[i]t may be that this is the only way the sexes can converse – when they're bodiless', might have come from Marinetti instead of a 1990s computer programmer (Cross 1995: 118–120). Both Futurism and cyberculture look to technology as a means of overcoming structured identities and differences. Both inadequately come to terms with the contradictions inherent in such a move, as we shall see.

The rejection of inwardness and depth goes well beyond the category of the aesthetic, however. This becomes evident when the Futurist approach to the subject is examined. Like Lyotard, Futurism's aesthetic theory is transposed onto the social realm itself. Futurism attempted to heal social and subjective divisions through reconstituting the subject and the social at a more abstract level. In contrast to the Symbolists who attempted to highlight such divisions and thereby question the ideals of bourgeois capitalist progressivism, Futurism openly celebrated the destruction of the self, through the harsh logic of the machine or through a kind of literary-cultural process of abstraction. Aesthetically, the aim is to destroy the literary 'I', the centred subject, implicated in depth psychology and bourgeois interiority. The Futurists declare that:

> [w]e systematically destroy the literary *I* in order to scatter it into the universal vibration and reach the point of expressing the infinitely small and the vibration of molecules . . . thus the poetry of cosmic forces supplants the poetry of the merely human.
>
> (Marinetti in Flint 1972: 98)

More broadly, the Futurists hoped that the destruction of the bourgeois 'I' would be made possible through technology. Ultimately, the Futurists aspired to 'the creation of a non-human type'. For them 'there is no essential difference between a human brain and a machine. It is mechanically more complicated, that is all' (Appolonio 1973: 8). Their focus on power and productivity led them to bypass the issue of what it means to be human, or the question of limits in relation to the technological reconstitution of everyday experience. They wished to abolish the categories of depth and inwardness, the province of bourgeois subjectivity, yet their desire to transcend the decadence of bourgeois life was full of hubris in its orientation.

The wish to destroy the bourgeois subject and to embrace a kind of post-human subject is not as simple as the Futurists would have us believe. First, it is clear that the Futurist attempt to move beyond the centred and rational subject of modernity masks a set of power relations that very much originate with the bourgeois (male) subject. For instance, the embrace of a machinic kind-of-being allows Marinetti to fantasise the possibility of male procreation. He writes of the urge to free oneself from the (female) body where we are 'sons and miserable slaves of the vulva'. Instead Marinetti proclaims that:

> [i]n the name of the human pride that we worship, I announce to you the coming hour in which men with wide temples and steel jaws will prodigiously give birth, only with their exorbitant willpower, to infallible giants . . . I announce to you that man's spirit is an untrained ovary . . . And that we are the first to fecundate it!
>
> (Marinetti in Blum 1996: 45)

The fantasy of male procreation and the flight from feminine materiality are not particularly novel. Rather, the impact of technological change allows these ideological fantasies to be rekindled. It is clear that some very human ideologies lie behind the wish to transcend the 'human' condition. The urge to engage in procreation without the female body serves as a particularised example of how nature is abstracted and reconstituted through the categories of Futurist culture. As we shall see, this desire for transcendence is also implicated in other metaphysical fantasies.

Embracing the transformation and empowerment of mankind through technology, Marinetti predicts a new kind of decentred subjectivity. He writes of 'Man multiplied by the machine. New mechanical sense, a fusion of instinct with the efficiency of motors and conquered forces' (Marinetti in Appolonio 1973: 97). We are faced with the by now familiar fusion of technological transcendence with more primal instincts. The certainty of these instincts is never questioned by the Futurists, despite their awareness of the changing nature of a human existence mediated by new technologies. Instead of a private, stable, interiorised subjectivity, we are given a multiple, public, technologised subject, freed from constraint, and able to roam the world freely. However, despite this gesture towards multiplicity, this subject remains predicated upon a basic essentialism; that of the primal or cosmic forces that will somehow provide orientation in the wake of the destruction of other certainties. Such essentialism can be read as a means of forestalling the ontological contradiction that ensues if the machinic and decentred mode of subjectivity becomes so dominant as to undermine the more centred and heroic sensibilities that motivate the initial desire for social and cultural transcendence. The attempt to recapture the 'certainty' of the primal in the absence of any other stable ground for subjectivity will feed directly into the glorification of violence. It is what will cause Benjamin to warn against any 'perverted attempt to re-enact the ancient ecstasies of cosmic experience' (Benjamin, W. 1986).

Indeed, rather than any cancellation of subjectivity, the Futurist fantasy of technological fusion reveals a new subjectivity, marked by a state of plenitude. The very concern with lack that motivates so much aggression towards women, in Futurist writing,[4] is solved through the constitution of a fantasised omnipotent subject, not so much annihilated, but rather reconstituted, through technology. Peter Nichols comments, on this new subject and the technological conditions of its production, that the 'machine abolishes the dialectic between inner and outer, private and public . . . [and institutes] a self whose thorough dehumanisation is the mark of its triumph over the lack and incompleteness associated with sexual difference' (Nichols 1989: 208).

Critics have noted how the apparent destruction of the self in reality masked the desire to fulfil the fantasy of subjective omnipotence. In particular, Blum notes that the supposed destruction of the 'I' 'does not involve the deconstruction of the unitary subject, but its expansion, its transformation into an all new powerful "I"'. She continues 'modern reality is not as in other modernist texts, a problematic world of contradictions to explore; rather, it is a world to be colonised and subjugated – fuel for a self-assertive, self-aggrandising enterprise' (Blum 1996: 41).

The Futurists destroy bourgeois subjectivity, but replace it by a technologically framed and more powerful being which engages with the world at a greater degree of abstraction, taking it apart and recomposing it at will. Engagement at this greater level of abstraction easily accommodates a violent, aggressive and subjectivist mode of relatedness to the world, whether it takes the form of straightening the Danube, male procreation, the creation of new poetry or the declaration of war.

The increasingly subject-centred mode of relatedness the Futurists advocated led to some obvious contradictions in terms of a Futurist 'revolution'. There is no doubt that the Futurists hoped to create some kind of broad revolution, indeed they often spoke of harnessing the energies of the masses. However, the Futurist revolt also often seemed to erase the possibility for a mass action or revolution. Despite their initial support for Russian Futurism and Gramsci's consequent praise, it soon became apparent that, in many ways the Futurist movement was antithetical to any kind of proletarian revolt. We can see that class conflict is negated by Futurist technology in two ways. First, Marinetti enacts a rhetorical strategy, which consists of displacing the worker's desires onto the emerging technology. He writes:

> Have you never seen a mechanic lovingly at work on the great powerful body of his locomotive? His is the minute, knowing tenderness of a lover caressing his adored woman . . .
>
> Of the great French railway strike . . . the organisers were unable to persuade a single mechanic to sabotage his locomotive . . . [yet] how could one of these men have been able to wound or kill his great faithful devoted mistress . . . his beautiful steel machine.
>
> (Marinetti in Flint 1972: 98)

For Marinetti, the worker experiences a powerful libidinal cathexis with the singular object of the (feminised) machine, which supplants any desire for collective revolt. It is clear that any revolution the Futurists anticipated would have to be individualised. Transcendence would come to those willing to transform themselves into a technologised way of being, rather than through any experience of class oppression.

Second, if technology allowed the subject a greater degree of control over the environment, if it limited the degree to which social, physical or cultural constraints could shape experience, then the result was an increasing atomisation of the masses, deprived of the conditions for a shared framework through which to interpret meaning. In other words, the technological will to abstraction embraced by the Futurists led to an increasing individualisation of the masses. This is reflected in Futurist rhetoric. In Marinetti's utopian society, he proclaims: 'Yes, the artists to power! The large proletariat of geniuses will govern' (Marinetti in Flint 1972: 98). Marinetti ignores the obvious contradiction involved in invoking a mass movement, based on a shared will for change, and the role of technology in dissolving the conditions which enable collective or shared experience. Instead, he individualises the terms of any possible 'revolution': in this new society the ability to stand outside social-cultural frameworks, the condition for genius, will be available for all. The masses undergo a process of atomisation: everyone governs and is an artist. The Futurist revolution then, is a particularly solipsistic one, as the conditions which would grant symbolic recognition to any revolutionary act fall away.

The Futurist body

If technology is sometimes constructed as a feminised object, it takes on distinctly masculine qualities when its possibilities are harnessed towards the construction of the Futurist body. Yet the Futurist body remains contradictory. On the one hand, the subject merges with technology to construct some sort of armoured body. The Futurist machine-man resembles Jünger's worker, in that both are detached from any close physical or emotional engagement with their environment. On the other hand, in 'The Tactilist Manifesto' (Flint 1972), the Futurists argued for a reclamation of bodily sensations, through a more rounded sensory experience than the sight-dominated tradition, and thereby implicitly suggested a non-phallic, dispersed model of sexuality which ought to run counter to the usual subtext of phallic aggression. Are these trajectories contradictory, or is something else at work here?

The Futurist body is one shielded from all the decadence of contemporary life. This new body arises through the 'imminent, inevitable identification of man with motor' and escapes contamination through subjecting the body's desires to 'metallic discipline'. Ultimately the futurist body will be 'a nonhuman type in whom moral suffering goodness of heart, affection and love, the sole corrosive poisons of inexhaustible vital energy, the sole interrupters of our powerful physiological electricity, will be abolished' (Flint 1972: 98).

The Futurist body functions as a kind of armour to protect the subject from what seem to be particularly 'feminine characteristics'. This defensive body is one wherein the subject is able to objectify itself. In an extraordinary image, Marinetti writes that 'on the day when man will be able to externalise his will and make it into a huge invisible arm, Dream and Desire, which are empty words today, will master and reign over space and time' (Flint 1972: 99). It is clear that Dream and Desire are regarded as empty precisely because they are structured through lack. The solution for the Futurists is to abolish lack through technological domination of the other. The technological body thus eliminates or controls anything alien to the self. Here, the Futurist body is constructed as a kind of proto-cyborg, marked by plenitude rather than lack, rejecting all inwardness and all embedded existence. Just as Mallarmé's 'depth' model was rejected in the aesthetic sphere, so here too we have an externalised will functioning as a huge arm, able to possess the world without fear of contamination. One is reminded of the Futurist joy in flying: there is a return here to the detachment of subject and object. The aggressive subtext of power and domination is at times even clearer. We are told that 'this nonhuman and mechanical being, constructed for an omnipresent velocity, will be naturally cruel, omniscient, and combative' (ibid.).

However, in the 'Tactilism' manifesto, written much later (1924), we see the emergence of a different perspective. For Marinetti, Tactilism's 'purpose must be, simply to achieve tactile harmonies and to contribute indirectly toward the perfection of spiritual communication between human beings through the epidermis' (Marinetti in Flint 1972: 119). Unlike the armoured and defensive cyborg body of the earlier works we are given a form of embodiment defined by closeness, which is not constructed through any kind of domination of the other. Indeed, the domination through the metaphor of sight, which we noted earlier, has been replaced in the following striking passage, by a new apocalyptic setting:

> Imagine the Sun leaving its orbit and forgetting the Earth! Darkness. Men stumbling around. Terror. Then, the birth of a vague sense of security and adjustment . . . men adapt themselves to the shadows . . . dimly explore the inside of their neighbour's bodies.
>
> (Marinetti in Flint 1972: 119)

Instead of penetration through sight, we have dim exploration: instead of a totalising vision, we have darkness. Rather than an omniscient will, there is uncertainty and an open permeation of the boundaries between subject and object. There is also an implied shift away from the model of genital sexuality. This shift is hinted at in Marinetti's wish to discover a plurality of senses: 'Today one can uncover and catalogue many other senses'; and in his insistence that Tactilism is directly opposed to 'morbid erotomania' (ibid.). In other words, the reclamation of tactility suggests a broader mode of sensuality than genital fixation.

It is tempting to regard the 'Tactilism' manifesto as some sort of corrective to the aggressively masculinist sensibility of the earlier writings. It is useful to recall the general horror of materiality that manifested itself in works such as the 'Destruction of Syntax' through to 'Man Multiplied by the Machine'. Indeed Marinetti's 'tactilism' would resonate with many who today advocate polymorphous embodiment, from Lyotard to some contemporary feminists.[5] However, two qualifications need to be made. The first is historical: Blum comments that '[t]he image of groping in the dark is a telling metaphor for Marinetti's bleak outlook on life at the time . . . of his compromise with the shifting politics of fascism' (Blum 1996: 134–135).

The second is more significant. While Marinetti's 'night vision' may suggest some form of openness to the other, it is clear that, for the main part, his project for a new tactility remains framed by the subjectivist metaphysics so evident earlier. For Marinetti, the world is still framed as a set of resources to be plundered by the subject. In the 'Tactilism' manifesto he writes that tactilism drives us toward 'the discovery of new senses' and that it allows us 'to penetrate deeper . . . into the true essence of matter' (Flint 1972: 120). It is possible to read the focus on tactility merely as a means of providing new experiences to be fetishised by the Futurist subject. Marinetti does not distinguish between the tactile body as providing a framework through which we open ourselves to the other, as David Levin theorises it (Levin 1995, also see Chapter 2 in this volume), and the body as a mere receptor for experiences framed entirely around the subject. Marinetti's model is perhaps closer to the mode of intersubjective relations constructed through virtual reality, where the subject can experience a range of sensations with fleeting others. Marinetti's model of a heterogeneous tactility serves as a particular example of a more general sensibility that attempts to liberate the subject from the body, from nature, and from the social and cultural other, through transcending the settings through which these have been constituted at, and engaging with them at a greater degree of abstraction. This level of abstraction allows the ideological fantasies of subjective mastery to unfold with a new vigour, as the constraints of the other are less binding. However, this is not to imply that this more abstract mode of engagement is unstructured. Indeed, the capitalist system of exchange provides the most logical framework within which the desires of Futurism could operate at their most unfettered.

Modern capitalism – a Futurist apotheosis?

There is no doubt that the Futurists wished to be liberated from the body in so far as the body served as a constraint to the desiring subject. As we have seen, this was to be achieved through the construction of a technological body, which kept out all forms of threatening otherness or through the reconstitution of the lived body as a series of multiple zones of sensual tactility. In both cases, we can see how embodiment is constituted more abstractly, either as an externalised body (the metaphor of the giant arm) able to possess the world unilaterally, or as a fragmentation of the body into a loosely interconnected series of tactile receptors. In both cases, liberation of the body comes through a liberation *from* the body, as it has been previously experienced. Andrew Hewitt argues that the ultimate fate of the Futurist body is to be *destroyed*, consumed in the flux and pure speed of commodity exchange (Hewitt in Goslan 1992: 50).[6] Paradoxically, this destruction is celebrated by the Futurists as the ultimate kind of liberation. Hewitt cites the following passage from Flint:

> The plainest, most violent of Futurist symbols comes to us from the far east. In Japan they carry on the strangest of trades: the sale of coal made from human bones. All their powderworks are engaged in producing a new explosive substance . . . This terrible new mixture has as its principle element coal made from human bones . . . For this reason countless Japanese merchants are thoroughly exploring the corpse-stuffed Manchurian battlefields. In great excitement they make huge excavations, enormous piles of skeletons multiply in every direction . . . One hundred *tsin* (7 kilograms) of bones brings in 92 kopeks.
> (Flint in Hewitt 1993: 50)

For Hewitt, the destruction of the body is the logical end-point of Futurist philosophy. The 'normal' body, which for the Futurists has become the site of 'bourgeois' desire and lack, of decadent emotions, fetishes and sentiment, of tangible otherness, is now transcended through its destruction and reconstitution as a transfigured object of pure exchange. Hewitt emphasises that this image of destruction does not equate with the Nazi genocidal destruction of Jewish bodies. Whereas Nazism seized on the concrete body of the Jew to avoid dealing with the abstract power of modern capitalism, this Futurist destruction willingly opens itself up to the abstract. If Nazism in this instance worked within a *restricted* economy (focusing on particular bodies), Italian Futurism harnesses a *general* economy of pure exchange. Destruction is celebrated to the extent that it reconstitutes what is destroyed within the circuit of pure abstract exchange. For Hewitt, burning the body,

> refuses the fetishisation of the body either as a static representation or as the finitude of exchange . . . the commodification of the body does not consist in its fixation and objectification but rather in the destruction of

any such fetish before the all consuming economy of exchange. The body is not the ultimate commodity, precisely because no commodity *can* foreclose the logic of exchange itself.

(Hewitt in Goslan 1992: 51)

Following this logic, Hewitt argues that, for the Futurists, 'self-destruction is not the opposite but rather the fulfilment of liberation' (Hewitt in Goslan 1992: 49).[7] Throughout this chapter, we have seen how the Futurists created a one-dimensional ontology, where depth, materiality, prior modes of being were all rejected in favour of a more abstract mode of engagement with the world that celebrates flux and exchange. Such a strategy, however, involves an ontological contradiction. The Futurist's desire for transcendence arises within the more concrete and less technological forms of early twentieth-century life. They look towards a more abstract form, enabled through technology, through which to achieve this transcendence. However, the conditions through which transcendence is achieved may actually erode the ground through which the transcendent act derives its meaning. In this case, subjective transcendence ultimately results in the destruction of the subject itself (Schnapp 1990).[8] A complete reconstitution within this level could only result in a meaningless Heideggerian 'functioning'. In the context of early-twentieth-century Europe, no such complete transformation was achievable. However, by affirming both physical and symbolic gestures of loss and destruction, Futurist discourse legitimates the fascist ritualisation of violence as an integral part of subjective and national renewal.

In this light we need to re-examine Marinetti's conception of tactilism. Its conditions of possibility, the night vision of tentatively groping subjects, can now be reinterpreted as a vision framed by the logic of pure exchange, where tactile impressions are detached from their originating subjects. The body disappears under the cover of darkness and arguably this is also the metaphoric condition where tangible, and identifiable intersubjective relations between subjects also disappear. What remains is only a disconnected and random exchange of abstract physical contacts. Critics have seen some sort of continuity between Futurism and postmodern nihilism (see Nichols 1989 and Blum 1996).[9] Certainly, both share an ontological ambivalence regarding the status of the other. If Futurism attempts to assimilate aggressively or destroy the other, then nihilism fails to distinguish between the self and the other, and consequently declares its impotence in the face of what it regards as a contemporary condition of overwhelming relativism. Yet, if we are able to reject some of the more nihilistic appraisals of our own postmodern era, it is worth remembering that technologies such as virtual reality are welcomed precisely for their capacity to blur the distinction between self and other, through the technological reconstitution of tactile sensations (see Chapter 7).

The commitment to a logic of pure exchange is further entrenched in the Futurists' understanding of time. In their founding manifesto, they aggressively reject the constitutive importance of time, boldly declaring that it 'died yesterday'. Extending his argument that Futurism aims for an apotheosis of pure exchange, Hewitt writes that 'Futurism aims to accelerate the temporality of modernity to the point where no moment of presence can be posited – except in retrospect as a despicable symptom of nostalgia' (Hewitt 1993: 109). Yet this is bound to result in another form of contradiction. Once temporality is framed through a productionist imperative, once the logic of progressivism is centralised, then the meaning of the *future* is diminished because it is always already domesticated through the logic of progressivism and innovation. The future as a temporal otherness is erased when all forms of temporality are collapsed into a circuit of pure exchange. If Futurism domesticates the present through a philosophical aporia, then capitalism enacts this domestication materially. As we shall see, Lyotard argues precisely this point in his paper 'Time Today'. The point here is that Futurism lays the philosophical grounds for a situation only the logic of commodity exchange can adequately fulfil. It is here that we do well to recall Benjamin's critique of undialectical theories of progress. Benjamin argues against the uncritical hypostatisation of progress where progress comes to stand for the course of history in its totality. Futurism's rejection of stasis, history, the past, the other, of depth and memory, produce this very kind of hypostatisation.

It is at this point that we can distinguish the particular fascistic impulses in Italian Futurism. Superficially, the Futurist worship of the machine, with its harsh 'metallic logic', its disciplinary properties, its violence and force, links the Futurists to fascism through the themes of discipline, power and control. There is no doubt that these aspects manifest themselves in Futurist discourse. Similarly, the man-machine hybrids constructed by Marinetti come to resemble the fortified fascist warriors in Jünger's writings. Yet it is the *other* side of the machine (remaining with Hewitt's schematic division) – the machinic properties of speed, movement, flux, and most importantly productivity – that most distinguish the proto-fascistic elements of Italian Futurism. Thus the Futurist merging of art and life means that every aspect of our being is reconstituted into a circuit of pure exchange, a relation that only modern forms of capitalism can accommodate. The destruction of older forms of rationality, order and logic – the rules of syntax, museums, the conventions of marriage, the barriers of time and space – were to be replaced by an even more powerful and totalising imperative, the abstract logic of modern capital. Ultimately, Futurism aimed at a 'relentless overcoming of resistance, a destruction of those differential and dialectical formations in culture which threatened to impede the homogenising movements of capital' (Nichols 1989: 210). While it is true that the abstract framework through which we engage with the world is easily harnessed to the logic of capital, the immediate context of Futurism provided the setting for the unfolding of Italian fascism.

At this point, while acknowledging the importance and influence of Andrew Hewitt's work for developing the ideas within this chapter, I wish to distance my own position. The major point of diversion lies in Hewitt's schematic division between a Futurist 'machine' that stands for order, domination and totalisation (fascism/later works) and a more progressive metaphoric machine that stands for transgression, productivity and liberation (radicalism/early works). Blum has already noted that such a scheme necessarily elides the fact that the metaphor of the machine *already* emphasised order, discipline and control in the early works, and to some extent Hewitt must ignore this (Blum 1996: 173 n. 66).

However, I would also question the invocation of a model that valorises transgression, production and liberation without sufficiently examining their conditions of possibility. I have already suggested that any form of liberation which arises through an abstract transcendence of the prior social forms in which the desire originated, faces the problem of ontological contradiction to the extent that the settings which grant significance to the transcendent act are also overcome. In the case of the Futurists, as with much of the discourse surrounding cyberculture, technology is celebrated for its ability to lift the subjects out of their prior frameworks of reference without giving sufficient recognition to this contradictory process. In addition to the dilemmas surrounding the Futurist mode of subject formation, we can identify another contradiction in the Futurist project for regenerating the nation-state.

The body, the nation-state and ontological contradiction

At first, the Futurist project for regenerating the Italian nation seems at odds with the remainder of their programme. Why is it that the Futurist imagination is so stamped with a distinct national identity? There seems to be a distinct tension between a sensibility that embraces abstraction and constructs a subjective mode of engagement that is global and rootless, and the continual support for national regeneration and national imperialism. If the Futurist sensibility is so powerfully nomadic, why is it that such an important role is reserved for the nation? Paul James has described how the nation, as a particular kind of abstract community, is itself framed in ontologically contradictory ways. He comments:

> [t]he nation and the nation-state have become extraordinarily important in framing the practices and subjectivities of contemporary social life from the individual's sense of identity to the 'collective activity of fighting wars'. At the same time, the increasing permeability of nation-state boundaries to waves of culture, capital and emigrants has provided the context for contradictory responses.
>
> (James 1996: xv)

The place of the nation-state within Futurist discourse provides an exemplary instance of this contradictory response. In relation to the nation, we can see how technology operates within different constitutive levels, and how the dominance of a given level either contributes to, or destabilises the nation. Technology plays a central role in making the nation possible. The technology of writing, in particular the newspaper, mediates the face-to-face integrative form in order to create, as Benedict Anderson writes 'that remarkable confidence of community in anonymity which is the hallmark of modern nations' (Anderson 1991: 36). With the Futurists however, who advocate a further degree of technological abstraction where the binding modality of the face-to-face is undermined, the desire to regenerate a sense of national identity sits uneasily beside the embrace of a technological Imaginary which dissolves the conditions under which any identity could be formed. We shall see how nationalism returns, however, as a violent 'return of the repressed'.

From the very first, Marinetti was careful to emphasise that 'it is from Italy that we launch into the world this manifesto of ours with its overpowering violence and incendiary powers' (see Marinetti in Flint 1972: 50). For the Futurists, the celebration of violence and war only makes 'sense' within a framework that supports the nation-state. As Marinetti wrote 'we profess a nationalism that is ultraviolent, anticlerical and antisocialist, an antitraditional nationalism founded on the inexhaustable vigour of the Italian blood' (Marinetti in Schnapp 1990: 62).

This vigorously nationalistic rhetoric took place despite a programme that consistently advocated the collapse of boundaries and desired the destruction of time and space itself. Schnapp has noted how Marinetti's orations consistently positioned the audience in relation to their particular national origin. Interestingly, he does not address his audience as potentially post-human subjects multiplied by the machine, but as distinct representatives of a national body. Schnapp remarks 'For Marinetti the audience is always the dynamic point of convergence between a particular national physiognomy or race, a national or cultural landscape, and a precise historical heritage whose imprint is in the blood' (Schnapp 1990: 63).

This invocation of nationality, tied as it is to the twin axes of physicality and memory, sits uneasily beside an overwhelming amount of Futurist writing that actively opposes memory and embodiment, within a framework that ought to cast the nation-state into obsolescence. Curious, too, is the continual invocation of nationalism through bodily metaphors such as 'Italian blood' and 'Italian democracy . . . a body that must be liberated'. The metaphor of blood, a symbol that traditionally draws upon more concretely framed experiences to unite a community of strangers, is used by Marinetti without any apparent sense of contradiction. There are other examples where Marinetti relies on the same metaphor to consolidate national identity. In the following example, strangely reminiscent of the tactilism manifesto, Marinetti is at his most comprehensive in emphasising the sheer *materiality* of the bodily metaphor:

> in its circular expansion the heart of man ruptures the suffocating familial circle, touching the extremities of the Fatherland, where it feels the pulse of border compatriots as if they were the outer nerves of its own body.
>
> (Marinetti in Schnapp 1990: 63)

We have the familiar desire for transcendence and liberation, but here it is curiously circumscribed by the limits of the nation-state. This phenomenon can be partially explained by returning to the metaphoric intertwining of the individual body and the nation-state. If we recall how the destruction of bourgeois subjectivity, whether by the machine or through the abolition of the 'I' in literature, masked the fantasy of an even more powerful subjectivity, which asserted itself through a violent eradication of all forms of otherness, then we can find a parallel process in this violent reassertion of nationalism. The subject occupies a similar structural position to the nation in Futurist literature. The conditions of possibility for both are erased by a technological framework that dissolves the spatio-temporal categories through which either could operate. In the case of the subject, the Futurists abolish memory, interiority, the body as an organic site of subjective meaning; in the case of the nation, they advocate the destruction of spatial categories, and the universalisation of experience through communications, media and subjective expansion. With both, then, there is an unacknowledged contradiction; both the subject and the nation-state continue to function despite the embrace of a framework that would abolish their conditions for possibility. Both reassert themselves, in this reconstituted context, through violence, domination and aggressivity. The machinic, multiplied, post-syntactical subject possesses the earth and appropriates the feminine; the nation-state exists and gains its identity only through the act of war. In this context, it is possible to identify the Futurists' relation to technology as inherently fascistic. By advocating a framework of radical abstraction made possible through technology, they commit themselves to the simultaneous abolition and regeneration of Italian culture and statehood. However, the erosion of the prior constitutive frameworks that granted meaning to subjectivity or nationhood, means the Futurists can only generate identity through violence and domination.[10]

Conclusion

The Futurists' infatuation with technology created the conditions whereby the subjectivist metaphysics, which Heidegger argued were the enabling condition of technology as *Gestell*, culminated in hubristic violence. Historically, this was the first time that humankind could, in any generalised material sense, transcend the parochial forms of social life and, through technology, reach out into the world for the material to construct social and cultural identity. Yet, even in the case of the Futurists, it is possible to argue that the meanings, anxieties and desires, which germinated within previous social forms, could not be so easily transcended in the more abstract mode constituted by technology. We only have to think of the Futurists' anxiety toward both women and nature. Such sensibilities were formed within a prior framework yet permeate their entire work. Similarly, we can consider the nation-state, which ought to be outmoded by a technological nomadism. It powerfully (if disastrously) reasserted itself within Futurist discourse. It is here that we need to return to the argument made in relation to Benjamin's 'Destructive Character'. The wish to transcend a disempowering or languid past can only make sense if the operation of the transcendent social form (enabled through technology) is to some degree held in check by previous forms. The Futurists attempted a one-sided destruction, resulting in a hysterical and violent reassertion of subjectivity and patriotism. These, I suggest, were the consequences of an attempt to realise a wholly technological mode of being in the world.

5 Between totalitarianism and heterogeneity
Lyotard and the postmodern condition

> One does not at all escape metaphysics by putting language everywhere
> (Lyotard 1971)

We have examined three approaches to technology in Heidegger, Benjamin and the Futurists. All three wrote when, for the first time, technology enabled the profound reconstitution of human ways of thinking and being in the world. In many ways, the Italian Futurists are the most straightforward in their approach, yet their concerns and general sensibility resonate right up until the present day, forming an important precursor for cyberculture movements. The Futurists, as we have seen, looked upon technology as a means of transcending what they regarded as the staid and decaying social and cultural forms of early twentieth-century life. The manner in which technology feeds already existing ideologies is clear in the case of the Futurists, especially Marinetti. Technology allows an extension of violent and patriarchal fantasies, culminating in a complicity with fascist politics. This much is well known. The first part of this text extended this analysis. First, I argued that the relationship between technology and Futurism as a cultural movement goes well beyond the empirical affinity for new technologies. The Futurists moved to embrace a more generalised sensibility of transcendence and material abstraction, a sensibility which framed their relation to 'non-technological' aspects of cultural and social life. Second, while technology allows the promulgation of violent fantasies, in that new weapons technologies allow for an aesthetics of destruction, this violence was also a 'logical' outcome of the resolution of various contradictions in relation to technology. The embrace of a technological framework in order to initiate a process of subjective and national regeneration also entailed erasing the prior framework from which both the subject and the nation-state drew their meaning. The technologically mediated subject and the nation-state in the age of technological globalisation were threatened as much as empowered. It is not surprising then, that their identities were restored only through a violent domination of otherness. Finally, it is important to register the fact that the Futurist embrace of a logic of flux and circulation, their love of speed and their revolt against depth or temporality set up the conditions for a particular sensibility that can be played out most effectively, if not through war, then through the logic of technologised forms of commodity exchange. It is worth keeping this in mind when we examine the discourses on cyber-technologies, where needs and desires are primarily consumated through the consumption of images and commodities.

Heidegger's analysis of technology is more comprehensive than that of the Futurists. By understanding technology as enabling a constitutive framework which shapes our mode of engagement with the world, he allows us to understand how a cultural movement such as the Futurists can represent the culmination of a subjectivist metaphysics. Heidegger's work causes us to think deeply about the conditions through which the possibilities of freedom and action might unfold. Yet, as we have seen, Heidegger's analysis is also limited. Politically, his affiliation with National Socialism can be contextualised, at least partly, as a response to the crisis of what he understood to be an impending technological nihilism. His later approach, which attempts to *suspend* the technological will-to-power, can be read as a response to his own attempt to influence the national will.

Nevertheless, I have argued that Heidegger's understanding of technology remains important and ought not simply be dismissed due to his political and ethical failures. As part of a more general argument, I have suggested how technology might be most usefully understood through the way it intersects with a variety of constitutive levels. Such an approach allows us to reject Heidegger's overdetermined portrait of technology as a global enframing, because it fails to consider the contradictory relationship *between* these constitutive frames. On the other hand, we can recognise the value of Heidegger's theory in so far as it provides a powerful analysis of the reconstituting capacity of technology.

With this in mind, I examined the work of Walter Benjamin, whose dialectical approach to technology can be strengthened through situating it in terms of the social forms carrying the meaning of technological mediation. Like the Futurists, Benjamin often approaches technology through the question of destruction. However, I have argued that such destruction, when endorsed, is more concerned with the *content* of a particular social form than with the form itself. Benjamin's project for harnessing new technologies to a radical politics can also be strengthened through this approach. Where Benjamin finds a radical potential in technology, such as when technology allows for the revealing and overcoming of such fetishised content, this ought not to come at the expense of the constitutive form which carries it. Otherwise, to use his own examples, stories become information, mimetic fragmentation degenerates into 'shock'. In addition, it became necessary to point out that Benjamin's project also concerned itself as much with conservation and recovery as with destruction. I have suggested that we are best able to negotiate a path between conservation and destruction through an approach based around the question of the broader constitutive forms that carry social and cultural meanings. An analysis of social life from the point of view of constitutive forms necessarily implies a distinction between such forms and their specific contents. In relation to Benjamin, I suggest that when he argues for the destruction of aura as being implicated in a historical system of violence and domination he does so at the level of *content* only. Moreover, if technology through various techniques allows for authentic modes of auratic experience to unfold, then it can do so only through existing in a reflexive relationship between the relatively concrete *forms* which allowed the original modality of aura to be revealed, and the more abstract forms constituted through technology.

It is here that we turn to more contemporary thinkers and cultural movements, concerning themselves with technology. Writing after the consequences of the Second World War, Lyotard is very much aware of the potential of technology to lend itself to, and increase the power of, a totalising mode of apprehending the world. Similarly, many of the discourses of cyberculture work within a dichotomy that contrasts totalisation against a more radical heterogeneity. Both Lyotard and those who wish to radicalise cyberculture turn to answers *within* the possibilities offered by new technology. At this point, it is worth recalling how Heidegger argued we could not solve the dilemmas of technology by technological means. Paul Virilio, whose work I also consider, would no doubt share Heidegger's pessimism. It is now time to examine the arguments for both sides and to attempt to find a way beyond them.

To some degree, Lyotard's work reflects upon and reacts to the rapid development in the last thirty years of what have come to be known as the 'techno-sciences'. The decline in metanarratives, the understanding of selves and social relations as constituted through language games, the commitment to heterogeneity, are all made possible by the freeing up of social bonds and forms of legitimation, itself made possible by technological change. In the final stage of his work, Lyotard focused more specifically on the question of technology and its relation to time, and to the limits of the 'human' in works such as *The Inhuman*. This chapter will argue that, despite the diversity of issues covered in Lyotard's vast body of work, it is possible to find a number of recurring themes. First, there is a continuing opposition of heterogeneity against totalisation and universality. Second, heterogeneity is achieved either through a process of *dematerialisation* – whether it be the flattening out of the 'organic' body into libidinal surfaces or the dissolution of communities into individuals functioning as nodal points in an abstracted network of exchange – or through a *figural* relationship – the point where an unrepresentable otherness works to disrupt the discursive regime that attempts to contain it, thus opening up a space of radical incommensurability. In Lyotard's work on judgement, the figural relationship is called a *différend*, nevertheless the structural relationship remains the same.

I will suggest that while Lyotard takes steps to mount a radical critique of technology through developing a strategy committed to undermining modern frameworks legitimated by reason and replacing them by looser, more heterogeneous arrangements, these attempts are ultimately insufficiently radical.

This is so because Lyotard commits himself to an understanding of social and cultural practices that take place only at a constitutively abstract level. This abstraction takes many forms: the replacement of communal relations with the temporary contract, the substitution of the tangible other with an unconstrained mode of subjective openness, the transformation of the organic body into a zone of libidinal surfaces, and the reconstitution of the modern subject as a postmodern self that exists as a point of interchange on the information network. These changes cause social meanings to operate in a fundamentally different register, and ultimately undermine Lyotard's own oppositional strategies. One of Lyotard's problems lies in the fact that he regards technology as problematic *only* when it facilitates the operation of grand or universalising narratives. This often causes him to advocate technological change, without sufficiently theorising the nature of this transformation and the contradictions that may ensue in this process.

This chapter will look at various aspects of Lyotard's work in order to demonstrate his commitment to a constitutively abstract mode of social being. I will examine how the valorisation of both heterogeneity and the figural manifests itself in Lyotard's writing on aesthetics and the body, and how these tropes are subsequently transposed onto the social realm. However, in the context of the information society, such tropes can only function in Lyotard's work though a process of *dematerialisation* – the dissolution of socio-historical constraints, and the transcendence of prior modes of engagement with the world. The disembodied level of social integration made possible through the increasing centrality of the information society, has, as I argued in Chapter 1, broken down older paradigms of rationality, and decentred the context through which cultural universals shape social practices. Lyotard is unequivocal in celebrating this transformation for the radical possibilities it creates. However, I will argue that this transformation is a more contradictory process than Lyotard acknowledges. At a more abstract level, where the contexts through which social and cultural meanings are acquired have become more decentred and open, the meaning of Lyotard's privileged terms, such as heterogeneity and liberation, need to be carefully examined.

Lyotard was among the first to link technological change explicitly with the 'postmodern condition'. He recognises the manner in which technology can powerfully reconstitute social relations. Somewhat like Heidegger, he foregrounds a dominant mode of engagement that takes place through technology (performativity) and recognises a need to adopt alternative strategies. Lyotard is neither a celebrant of technocratic ideals, nor does he succumb to Baudrillardian resignation. He is a philosopher who is at times seriously concerned with questions of justice, one of the most political of the postmodernists, a supporter of the local against the universal, with a passionate commitment to the recognition of difference. As such, his work is worth examining for an alternative to the current hegemonic intertwining of capital and technology. The question remains as to whether Lyotard's conceptual framework, which attempts to grapple with and respond to technological change, is capable of realising his goals of justice and heterogeneity.

Before these issues are canvassed in relation to Lyotard's theory of postmodernity, I want to examine the opening of an earlier work: Lyotard's *Libidinal Economy*, so as to understand his commitment to a mode of engagement that is constitutively abstract, referring in this case to our modes of embodied experience.

Libidinal economy, the body and constitutive abstraction

The opening of Lyotard's *Libidinal Economy* is one of the most striking in modern philosophy and is worth quoting at some length. Lyotard invites us to:

> [o]pen the so-called body and spread out all its surfaces: not only the skin with each of its folds, wrinkles, scars, with its great velvety planes, and contiguous to that, the scalp and its mane of hair, the tender pubic fur, nipples, nails, hard transparent skin under the heel, the light frills of the eyelids, set with lashes – but open and spread, expose the labia majora, but also the labia minora with their blue network bathed in mucus, dilate the diaphram of the anal sphincter, longitudinally cut and flatten out the black conduit of the rectum . . . open the trachea and make it the skeleton of a boat under construction . . .
>
> (Lyotard 1993: 1)

This proceeds at some length, with Lyotard dissecting each part of the body and flattening it out until only a surface remains. The aim is to transform the organic body into a network of libidinal surfaces. Lyotard urges us to 'go immediately to the very limits of cruelty, perform the dissection of polymorphous perversion, spread out the immense membrane of the libidinal body' (Lyotard 1993: 2). The notion of the body as a site which could ground individual and communal relations is rejected in favour of this fragmented and 'abstracted' body.[1] We are presented with the body as a site of heteronomous desires, which can be put together and taken apart more or less at random. In *Driftworks*, Lyotard makes this more explicit, arguing that 'there has never been anything but pieces of the body and there will never be a body' (Lyotard 1984a: 10). He thus rejects any more concrete relation to the body as a relatively stable signifier intertwined within a specific range of social practices. Rather, Lyotard's claims about the body are representative of the transformation that I pointed to in the introduction where the social meanings of embodiment have become more arbitrarily constituted. Indeed the body itself is no longer regarded as a stable symbol of identity, but is subject to a variety of practices which can reconfigure its meanings. Lyotard's conception of the body as 'in pieces' serves as a particular example of a more general transformation.

While *Libidinal Economy* is notable for its rhetorical excess, more sympathetic critics have tried to praise it for re-introducing the notion of the body into philosophy (Sim 1996: 21). Yet, as we have seen, this is a body constituted at a relatively abstract level and its meanings constituted in an arbitrary manner. The socio-cultural settings that give the body its meaning are bypassed in this new framework, which promotes the desire for a series of discontinuous intensities that take place on the surface of the body. This can be contrasted to a less abstract mode of embodiment, where as we have seen in Chapter 1, the meanings of the body are inscribed within a specific range of social practices. At this level the body functions as a site of condensed meaning structured by relatively stable social and cultural settings. Yet, as we shall see, Lyotard also wishes to transcend these settings.

Postmodern subjects must take advantage of their heightened conditions of autonomy and 'seize every chance to function as good intensity-conducting bodies' (Lyotard 1993: 262). This purely subjective experience occurs in the absence of the social settings that would normally help to ground embodied experience. Perhaps this imperative is taken to its most extreme form when Lyotard claims that industrial injuries, such as deafness, are in actuality sites of *jouissance* for workers, because they entail 'the liberation of sensual potentialities from the rule of the integration of experience into the economy of the organic body' (Lyotard 1993: 120). Elsewhere Lyotard takes this notion further, sounding at times like Jünger who similarly aestheticised technologically-mediated experience in relation to the 'worker'. For Lyotard:

> there is in the worst conditions of the worker, a startling contribution, a contribution which can easily be paralleled to the adventure of poets, painters, musicians . . . [a contribution] to disrupting the measure of the human condition, to a putting up with situations which one might have considered unbearable. It demands another body, in another space . . . In particular an experience of quantity [intensity] without precedent in the rural tradition.
>
> (Lyotard 1990: 22)

In both cases, we have Lyotard advocating the 'mass destruction of the organic body' (Lyotard 1993: 111), it seems at any cost. We can see how the obsession with the overthrow of rationality and solidity in favour of a heteronomous field of libidinal intensities, which dominates *Libidinal Economy*, also informs the work on postmodernism and technology. First, Lyotard's stress on the incredulity towards narratives derives from a recognition that 'libidinal economy always and inevitably, disrupts the desired union of theory and practice . . . [it] calls into question all efforts of grand narrative closure' (Sim 1996: 40). Indeed, the organic body functions like a grand narrative, subordinating heterogeneous impulses within an ordered framework. It is clear that Lyotard wishes to overthrow the hegemony of the organic body. Yet the question remains: what meaning can embodied experience have if cast within settings that allow for the social meanings of the body to become purely arbitrary? In such a context, to what extent can one speak of a 'body' at all?

These tensions resurface in his essay, 'Can Thought go on without a Body?' (1991), where Lyotard examines the finitude of the earth, the end of the species and the role of the techno-sciences in relation to these concerns. The essay is divided into two sections, 'He' and 'She'. 'He' takes a basically pro-development position, arguing that with the end of the sun 'there will arrive the demise of your phenomenology and your utopian politics' (Lyotard 1991: 9). 'He' argues that, in this context, philosophy and politics are irrelevant, if not downright useless. The only real task for scientific thought is 'how to make thought without a body possible' (Lyotard 1991: 13). Clearly, such a meta-narrative of finitude and the preservation of thought for its own sake overrides any commitment to the operation of little narratives. At this point, 'She' responds for the second part of the essay. 'She' reintroduces the question of the body, in terms of the necessity of suffering to thought, and the irreducibility of gender difference and its essential relation to thought. There is even speculation that gender difference might also be ontological difference, thus constituting the ultimate *différend*. 'She' argues that human thinking is not simply an activity that can be replaced by machines. Opposing the primary goal of technological development, she argues, very much in a Heideggerian idiom, that the mind isn't 'directed' but rather 'suspended' – 'you teach it to receive'.

There is much to be commended in such sentiments. In this essay, we have Lyotard adopting a much more ethical tone, reintroducing the question of embodiment without the nihilism of *Libidinal Economy*. Indeed, in this essay the body plays an essential role in opposing the performative principle that frames techno-scientific development. Elsewhere too, we find Lyotard promoting the body as a locus of resistance:

> the body is to my mind an essential site of resistance, because with the body there is love, a certain presence of the past, a capacity to reflect, *singularity* if this body is attacked, by techno-science, then that site of resistance can be attacked. What is the unconscious of a child engendered *in vitro* ? What is its relationship with the mother, and with the father . . .
> (Lyotard 1986: 30–31)

Yet what type of body is able to constitute such resistance? Is this the dissected body existing as a site of libidinal surfaces? Is this the body of the deaf labourer, experiencing the technological destruction of the organic body as *jouissance*? On the contrary, this is the body existing in a 'depth' relation to the social realm that surrounds it. In other words, a body constituted at a less abstract level. Yet Lyotard never discusses the complex social settings that give the body a meaning. Thus, in *The Inhuman* all he can do, in the face of the total reconstruction or even elimination of the body by rapid advancements in techno-science, is to (like Virilio) suggest that we 'mourn' it. While Lyotard might have brought philosophy back to the body, it is hard to see how such a body could ever function as a site of resistance. As we shall see,

Lyotard's commitment to the dematerialising structure of the information society compels him to advocate the former kind of abstracted body. The body that exists only as the site of intensities is ultimately a reified body, which can play little part in complementing a lived sense of personal being.

The abstraction of the body in *Libidinal Economy* coincides with the abstraction of social relations in other works. Lyotard embraces the transience of individual encounters, over the notion of more concrete forms of communal relation. What happens when the 'libidinal' economy is transposed onto the social realm? Lyotard writes, in a passage that resembles Marinetti's description of nocturnal tactility, that: 'there are only encounters, each tracing at full speed around itself a multitude of transparent walls, secret thresholds, open grounds, empty spaces in which each encounter flees from itself' (Lyotard 1993: 36). Furthermore 'there is nothing permanent from one encounter to another' (Lyotard 1993: 38). Thus, for Lyotard, the face-to-face exists only in the more abstract mode of an 'encounter', rather than a binding *relation* framed by co-presence. In his attempt to escape rationality, representation and universality, Lyotard provides us with transience, flexibility and plurality. Yet, as we have seen, such a move comes only through embracing abstraction, as selves and social relations are lifted out from the settings which once sustained them. This becomes more explicit when Lyotard examines the possibilities of the information society.

The postmodern condition and the techno-sciences

One of the meanings of the 'postmodern condition'[2] is as a periodising concept which looks at the state of knowledge after the impact of the techno-sciences, in particular those which concern themselves with information technology. Lyotard's work is concerned with how new technologies can reconstruct the pattern of social relations. According to Mark Poster, 'Lyotard explicitly draws connections between language and electronic mediations on the one hand, and the contemporary social context on the other hand' (Poster 1990: 141). For Lyotard, the 'impact' of technology ranges across areas as diverse as communications, aesthetics, the construction of subjectivity and social relations, and leads to the necessity of rethinking the issues of justice and politics. However, the major change for Lyotard lies in the decline of what he calls 'grand narratives'. Modernity is marked, according to Lyotard, by its dependence upon universalising narratives. Such narratives ground legitimation: local phenomena are understood by reference to a universal framework. Lyotard examines the two species of grand narrative that have governed the modern era. The first is the 'narrative of emancipation' (political), the second the 'speculative narrative' (philosophical). Lyotard's critique of grand narratives is largely a critique of the modern state which relies on them to wield power. Thus '[t]he state resorts to the narrative of freedom every time it assumes direct control of the "people" in the name of the "nation", in order to point them down the path of progress' (Lyotard 1984b: 42). Lyotard claims that universalising narratives lead to totalitarian possibilities, fascism and Stalinism, for instance. With respect to technology, he advocates the changes it brings precisely in so far as these break down grand narratives. Throughout his work, Lyotard presents us with a stark choice, between the terror of universality and the heterogeneous possibilities contained in the local. As we shall see, this leads to problems when Lyotard examines the information society, which is itself a *universalising* process.

Lyotard's central question in *The Postmodern Condition* asks: 'where after the metanarratives, can legitimacy reside?' His hope lies in the proliferation of the local, made possible after the breakup of universal frameworks. Lyotard recognises that, while technological change can lead to a greater degree of terror, the profound changes that it brings about also lead to the emancipatory possibility of decentralisation, a mode of engagement that Lyotard affirms across all spheres. The rise of the information society leads to the increasing irrelevance of the nation-state, since information technologies allow relations and transactions to transcend political and geographical boundaries.

Of course, it is not just in the area of computerisation that these radical changes have manifested themselves. Lyotard notes how techno-science has the very real potential to reconstitute matter itself. The rise of new technologies such as genetic engineering promise a 'profound transformation of the relationship between man and nature' (Lyotard 1986: 10). These changes allow for a celebration of the possibilities of new freedoms, which arise through the breaking up of older constraints. The linear progression of history, the unifying narrative of reason, the idea of 'man' as the centre of knowledge, are all made problematic in the postmodern era. Instead we have a more relativistic model of 'local' knowledges and communities where, according to Lyotard, the idea of a unifying, totalising form of legitimation would seem an arbitrary form of terror.

Certainly there is much to be said for decentralisation, particularly in the light of the new social movements, the rise of postcolonialism and so on, all of which problematise the 'modern' mode of legitimacy, predicated on Eurocentric, masculine norms. Indeed, the idea of pluralism sounds appealing in a context where, as Feenberg acknowledges:

> the notion that there are multiple rationalities, that modern science is not alone in understanding the universe, is intended to level the playing field, to allow for difference and variety in a way that the old positivist faith in science would not.
>
> (Feenberg 1995: 131)

However, the situation is a little more complicated than Feenberg suggests. While such 'levelling' might superficially allow a space for difference to emerge, it poses the question of how difference is to be accommodated within this new framework. What conditions would allow the subject to respond to the complexities and contradictions that revolve around the issue of difference, in a way that might lead to a meaningful sense of engagement with others in the lifeworld? What might be the significance of difference for the subject constituted in Lyotard's version of postmodernity? How could difference operate, in the absence of all universal frameworks in ways that would not lead to pure relativism? Whether Lyotard is able to respond to these questions is something that we shall have to return to later.

By way of orientation, I would like to separate my own position from Lyotard's at this point. We can distinguish between a pluralism based upon the notion of intersecting constitutive forms, and a pluralism of *content*, constituted within a single abstract level. The former argument is the one made by this book. It argues that less abstract forms, which provide contexts of certainty and stability, ought to co-exist with more abstract forms which allow for a plurality of social practices. By contrast, Lyotard's approach remains fixed on a plurality of content, and is indeed predicated upon the transcendence of prior forms. However, we need to examine his work in more detail, before the contradictory nature of an argument for plurality contained within a single level can be revealed.

While Lyotard is concerned with the broad changes technology brings about, in *The Postmodern Condition* he concentrates for the most part on information technologies. The rise of information technology is symbolised by the computer. Lyotard examines the relation between computers and the process of knowledge gathering, noting a profound transformation. The computer marks the end of the humanistic ideal of knowledge as the self-construction of the subject (Feenberg 1995: 12). Lyotard writes that:

> Along with the hegemony of computers comes a certain logic, and therefore a certain set of prescriptions determining which statements are accepted as 'knowledge' statements . . . We may thus expect a thorough exteriorisation of knowledge with respect to the 'knower', at whatever point he or she may occupy in the knowledgable process. The old principle that the acquisition of knowledge is indissociable from the training (*Bildung*) of minds, or even of individuals, is becoming obsolete, and will become even more so . . . Knowledge is and will be produced in order to be sold, it is and will be consumed in order to be valorised in new production: in both cases the goal is exchange. Knowledge ceases to be an end in itself, it loses its 'use-value'.
>
> (Lyotard 1984b: 4)

This new status for knowledge leads to many questions. If all knowledge is 'flattened out', functioning only within the commodity form, where does meaning reside? What, in this new social terrain would make it possible to act differently towards a future that would differentiate itself radically from the modern? For Lyotard, the situation is not necessarily cause for despair. On the contrary, the process of dematerialisation, the destruction of universal narratives, the potential to alter every aspect of our being through technology, frees up the social realm in a manner previously unthinkable, and allows new strategies based around heterogeneous social relations. The basis for such relations comes from a specific kind of aesthetics transposed onto the social. Clearly, for Lyotard, the possibilities embodied within technological change occur through re-igniting the avant-garde tradition and through re-theorising the sublime. Yet as we shall see this conception of the aesthetic cuts right across the social. Fredric Jameson remarks that:

> Lyotard's aesthetic positions, however, cannot be adequately evaluated in aesthetic terms, since what informs them is an essentially social and political conception of a new social beyond classical capitalism . . . the vision of a regenerated modernism is, in that sense, inseparable from a certain prophetic faith in the possibilities and promise of the new society itself in full emergence.
>
> (Jameson 1992: 60)

The promise of this 'new social' is predicated upon the breakup of older frameworks. Technology is a major cause of this dissolution, as information changes the character of knowledge, genetic engineering changes what it means to be 'human' and the breakdown of grand narratives commits us to legitimate the social in new ways. However, the radicality of Lyotard's break with prior forms of being might lead to an even greater threat of terror than that contained in the modern. This issue can only be dealt with after Lyotard's alternative framework is considered in more detail.

Technology as totalising – performativity versus paralogy

The dominant framework through which technology operates is clearly not one which Lyotard supports. He argues that the rise of the 'information society' threatens a new degree of totalisation, embodied in the culture of 'performativity'. Performativity is the dominant mode of technocratic thinking that occurs when technology fuses with the rule of capital. In a culture of performativity, Lyotard argues, 'knowledge and the social order can be legitimated simply by the continuous perfecting of means without appeal to either traditional symbols or modern ideals' (Feenberg 1995: 130). Here it is possible to compare Lyotard's thinking of technology with Heidegger's concept of the *Gestell*, in that both consider the impact of technology in relation to the constitutive frame through which it occurs. On this point, Poster writes that 'computer technology, Lyotard reminds us in the spirit of Heidegger, only comes into play in a general social context in which performativity is already the master code' (Poster 1990: 141). For Lyotard, the danger of the current 'master code' is its tendency towards the ideals of system control and totalisation: the elimination of contingency through context control marks the danger of a generalised computerisation of society. He notes that:

> the performativity of an utterance ... increases proportionally to the amount of information about its referent one has at one's disposal. Thus the growth of power, and its self-legitimation, are now taking the route of data storage and accessibility, and the operativity of information.
> (Lyotard 1984b: 47)

Hence, the breakdown of older modes of legitimation does not necessarily lead to emancipation. On the contrary, computers pose the threat of a cybernetic society of total administration. How does Lyotard respond to this? Broadly on two fronts. First, he argues that contemporary science demonstrates that totalisation is an impossible goal. Any system will eventually be undermined by the forces of heterogeneity and chaos within it. Consequently, the findings of a new science could generate an alternative model of legitimation to that of performativity. Second, and more significantly, Lyotard advocates new strategies for resistance, strategies made possible by the technological reconstitution of the social realm. The freeing up of social relations, the decentring of the subject, and the abstract conception of the social as a network consisting of different 'language games' allow alternative modes of legitimation and alternative modes of being to arise. Lyotard will claim that such modes are able to resist the threat of totalisation, through acting within an aesthetic framework which privileges heteronomy and difference.

Lyotard claims that technology is marked by a certain ambivalence. On the one hand, it could serve as the 'dream' element in a totally administered society. Yet Lyotard claims that such totalisation is an impossible end, that it is undercut from within. In *The Postmodern Condition*, he claims that the main argument against any form of legitimation by performativity arises through modern science. Chaos theory, for instance, renders the technocratic dream of absolute control impossible. The epistemological shift in science is one that necessarily accepts chaos and randomness, and which refuses all claims to absolute knowledge or control. According to Lyotard:

> Postmodern science ... is theorising its own evolution as discontinuous, catastrophic, nonrectifiable, and paradoxical. It is changing the meaning of the word knowledge, while expressing how such a change can take place. It is provoking not the known but the unknown.
> (Lyotard 1984b: 60)

Lyotard claims that such findings may provide an 'alternative model of legitimation', based in the continual overturning of paradigms, placing an emphasis upon the local instead of the universal, and promoting an ideal of heteronomy and openness.

While the spirit of Lyotard's argument might be impressive, it is possible to raise several objections to his reliance on the cognitive destabilisations of contemporary science. Seyla Benhabib has argued that Lyotard fails to distinguish between the internal cognitive dynamics of contemporary science and its social uses (Benhabib 1984: 114). While Lyotard privileges the internal epistemological dilemmas that arise within scientific theory, he pays insufficient attention to the socio-technological aspects of scientific practice. In other words, how could chaos theory provide an alternative model for legitimation, when it is swallowed up in a scientific practice dominated by performativity? Lyotard fails to consider such issues.

Second, it is possible to question Lyotard's assessment of scientific theory itself. While he argues that such theory will provide a model of randomness, instability and heterogeneity that will privilege the local over the universal, some commentators claim that this is not necessarily the case. For instance, N. Katherine Hayles argues that the new science, to a large extent, still frames its theory through reference to the global and the universal. She writes that:

> [t]he endorsement of local knowledge within contemporary science cannot be the panacea that Lyotard imagines. It is true that the new paradigms recognise the importance of scale, and therefore of locality. But it is also true that these changes are located within disciplines committed to universal theory . . . [a]s these scientists use 'chaos,' it connotes not the unpredictable aspects of disordered systems but their universal characteristics. To see chaos theory as the antidote to totalisation is to ignore this thrust toward universality.
>
> (Hayles 1990: 215–216)

However, while these critiques may reveal the inadequacy of Lyotard's reliance on the model of scientific theory for a new mode of legitimation, they do not substantially damage his claims that heterogeneous social relations are made possible by new *technologies*. Such technologies have, at the very least, the power to bring into question modern modes of legitimation, even if they cannot by themselves suggest new ones. As we have seen, the capacities of technology to reach across time and space, to break down older modes of being, and to radically bring into question almost every taken for granted notion of human existence leave the possibility of a radically new future that differs significantly from the modern as an open question. It is time to consider the various strategies Lyotard would advocate for such a future.

Strategies

Lyotard sees two possible alternatives under postmodernity. The first leads towards totalisation. The prevailing culture of performativity corresponds with the threat of a form of terror, the pursuit of an absolute system. For Lyotard, such a universalising practice could only deny difference and freedom. The alternative is a practice that allows for the unfolding of heterogeneity. Lyotard poses the issue thus:

> The system, as it exists, absorbs every consistent discourse: the important

thing is not to produce a consistent discourse but rather to produce 'figures' within that reality. The problem is to endure the anguish of maintaining reality in a state of suspicion through direct practices; just like, for example, a poet is a man in a position to hold language – even if he uses it under suspicion.

(Lyotard 1984a: 79)

In *The Postmodern Condition* this strategy revolves around paralogy and the cultivation of 'little narratives'. As we shall see, these strategies rely on a specific conception of subjects and social relations in order to operate. But first to their specifics. Paralogy informs a new paradigm emanating from contemporary science, which posits a mode of legitimation based upon 'the continual search for new moves which challenge the consensus of dominant paradigms'. It is marked by the search for innovation, the overturning of past certainties, and the refusal to be subordinated to any universal 'theory'.

Following the practice of contemporary science, Lyotard claims that the social realm could be constituted through the operation of a multiplicity of 'little narratives'. Lyotard compares such little narratives with the narrative modes of primitive societies. In so doing, he expressly bypasses the modern period and its concern with metanarrative. Sim argues that this 'can be seen as an expression of that commitment to a dialogue with the past, which has as its motivation the desire to undermine the authority of modernity/modernism' (Sim 1996: 49).

Lyotard argues that narrative is the primary mode of knowledge in pre-modern societies. Popular narratives are self-legitimising, they do not require justification by way of a metanarrative. Such 'small' narratives are legitimated pragmatically, not by way of argument and logic, but by practical use of the 'know-how' they embody. The emphasis on the pragmatic and performative elements of these modes of narrative is posited against legitimation by metanarrative *and* against the dominant mode of performativity. On this point, Ingram remarks that 'the cybernetic model overlooks this fact when it attempts to reduce communication to the tranquil exchange of information within a closed system' (Ingram 1987: 291).

Lyotard claims that 'the little narrative remains the quintessential form of narrative invention' (Lyotard 1984b: 60). Unlike grand narratives, which are seen to force closure by reference to some unifying principle, little narratives provide the means to carry on the scientific practice of paralogy within a broader social sphere. According to Bill Readings, little narratives promote openness and flexibility, creating new spaces for innovation because, unlike grand narratives,

> each little narrative does not aim to tell *the* story, to put an end to narrative; rather a little narrative evokes new stories by the manner in which in its turn it has displaced preceding narrative in telling *a* story.
>
> (Readings 1991: 60)

We have a sort of radical positivism, a mode of engagement that deals with tradition, not by remaining within its dominant paradigm, but by reacting against it. Yet, while this process may seem appealing in that it advocates a kind of continual revolution, we need to examine more closely the conditions of possibility that would enable this radical overturning to be significant, and not end up a mere variation upon Heideggerian 'functioning'.

One of the reasons that traditional narratives are privileged over grand narratives is because they are self-legitimating. They do not refer to an 'outside'. How, then, does legitimation occur? For Lyotard, such narratives are legitimated because 'they do what they do', in other words they are *internalised*. In pre-modern cultures, narratives function through ritual and custom. The narrator is subservient to the tale. Such pragmatism ensures that the question of legitimation does not even arise. Yet the same cannot be said for 'small' narratives in the context of the information society. While Lyotard does not attempt to force any absolute affinity between pre-modern and postmodern modes of narrative knowledge, he does attempt to find some similarity between them. In Chapter 1 I argued that it was crucial to recognise that the meaning of any social practice was related to the broader constitutive form that carried it. Lyotard's 'small' narratives are a case in point. 'Little' narratives constituted at the face-to-face level of social integration are qualitatively different when constituted within a social formation where the level of disembodied integration is dominant. If, in the case of traditional narratives, legitimation did not come from 'outside', it was nevertheless embedded in face-to-face relationships of tribal or communal societies. 'Little' narratives are of a quantitively different order. If they are not legitimated through reference to grand narratives, they are not legitimated through a tightly-bound social formation either. On the contrary, postmodern social relations are, as Lyotard continually reminds us, notable for their degree of openness and their transient nature. It is worth examining two issues here: the pragmatic element contained within 'little' narratives, and the conditions that allow for 'little' narratives to operate in the information society.

The pragmatic element Lyotard emphasises in little narratives works to alter profoundly the *value* that gets assigned to differing modes of knowledge, as they are disentangled from their prior frameworks of reference. This leaves the problem of how we are to interpret different kinds of knowledge. In a universe governed by a multiplicity of little narratives, '[v]alue does become a rather arbitrary quality under such a dispensation (the criteria for a discourse being whether it "works" or not)' (Sim 1996: 45). Given this assessment, Lyotard's position appears closer to the principle of performativity than he might desire.

Some critics also claim that the dissolution of interpretive ground, as grand narratives split off into a proliferation of self-contained little narratives, causes a loss of critical distance. Christopher Norris is concerned that:

> [a]s the idea gains ground that *all* theory is a species of sublimated narrative, so doubts emerge about the very possibility of *knowledge* as distinct

from the various forms of narrative gratification. Theory supposes critical distance between its own categories and those of naturalised mythology or commonsense set of assumptions. Simply to collapse that distance – as Lyotard does – is to argue away the very grounds of rational critique.

(Norris 1995: 23)

Norris is right to point out the lack of critical distance that arises in the absence of universal theories. To date, Lyotard has not solved the question of justice in the postmodern era.[3] At no time does he consider the possibility that the local and the universal might exist in some sort of reflexive relationship. On the contrary, technology is regarded as valuable to the extent that it is able to undermine universal narratives, and dangerous when it enshrines them. Yet, as we shall see, the information society itself creates a universalising process that radically undermines the meaning of the 'local' in a postmodern age.

Subjectivity and social relations in the information society: Lyotard's postmodern condition

Following Meaghan Morris, it is possible to detect a continuity in Lyotard's work, in terms of the constant privileging of formal experiment and rhetorical difference, whether this be in aesthetics, social relations or politics (Morris 1985: 50). It is important to note that, in the theory of postmodernism, Lyotard's aesthetic commitment to abstraction is transposed on to the social, and that *it is the information revolution that allows him to do this*. This section will examine the way Lyotard constitutes subjectivity and social relations in the information age, and will then proceed to outline his aesthetic theory, so as to show how it informs his stance on technology. We can recall that in *Libidinal Economy*, Lyotard rejected the notion of the organic body and the concept of bounded social relations. This assertion returns and is made more explicit, since the computer revolution constructs both subjects and social relations via a linguistic framework. As Kroker notes 'Lyotard's world *actually* begins with the construction of subjectivity by the norms of technology' (Kroker 1992: 150). In *The Postmodern Condition*, Lyotard makes this explicit, arguing that:

> [a] *self* does not amount to much, but no self is an island; each exists in a fabric of relations that is now more complex and mobile than ever before . . . a person is always located at 'nodal points' of specific communication circuits, however tiny these may be. Or better: one is always located at a post through which various kinds of messages pass. No one, not even the least privileged among us, is entirely powerless over the messages that traverse and position him at the post of sender, addressee, or referent.
> (Lyotard 1984b: 15)

Here we can see that the fabric of relations that structures the self has become dematerialised. Through communication circuits, prior modes of engagement with the world are reconstituted in a more abstract setting. While the self may not amount to much, Lyotard implies that is it is in some way empowered. 'Even the least privileged' are able to shape themselves in this new setting. Yet while this reconstruction of subjectivity is celebrated for its apparent freedoms, this is a freedom established only in the terms allowed for by the new framework of information exchange. Lorenzo Simpson sees this medium of networked exchanges as contributing to a new form of Cartesianism manifested in the information society. He claims that: '[j]ust as the subject's very being, for Descartes, was hostage to thought, for the postmoderns it is hostage to information exchange – I transmit and receive, therefore I am' (Simpson 1995: 213). Lyotard, however, argues that this new situation allows the subject a greater degree of emancipatory potential, which arises through the dissolution of older constraints. He claims that:

> this is what the postmodern world is all about. Many people have lost the nostalgia for the lost narrative. It in no way follows that they are reduced to barbarity. What saves them from it is the knowledge that legitimation can only spring from their own linguistic practice and communication interaction.
> (Lyotard 1984b: 41)

Here we can see Lyotard taking on a broadly post-structuralist conception of subjectivity, where the subject is constituted through a multiplicity of linguistic practices. The possibility for legitimation lies in the subjective ability to transcend older paradigms of being and to recreate oneself autonomously through technology. Significantly, the subject becomes responsible for its own meaning, recreating itself within communication circuits of exchange. This is the hope that lies behind Lyotard's call, towards the end of *The Postmodern Condition*, to give the public free access to the databanks. While access to information is an important issue for straightforward political reasons, Lyotard also seems to be referring to the processes of self-creation. The implication is that, the more information at hand, the more enriched the process of subjective engagement with the world. This marks an important transformation. Within modernity, subjective creation was to some extent grounded by social and cultural constraints as I pointed out in Chapter 1. This dominant mode of subjective engagement is, within postmodernity, replaced by an autonomous form of free-floating subjectivity. Lyotard combines Wittgenstein with a post-structuralist flavour to describe this transformation. He writes that:

> the *social subject itself seems to dissolve* in this dissemination of language games. The social bond is linguistic, but is not woven with a single thread. It is a fabric formed by the intersection of at least two (and in reality an indeterminate number of) language games obeying different rules.
> (Lyotard 1984b: 40, emphasis added)

The emphasis, as with much of post-marxism (for example, Laclau and Mouffe) is on the possibilities enabled by a radically *open* mode of social integration. In Lyotard's case, this openness is established by regarding the social realm as a 'sort of infinite field of linguistic possibilities'. According to Lyotard, social relations are already increasingly governed by what he labels the 'temporary contract'. The temporary contract describes the new form of interaction that dominates postmodernity. It emphasises the local and fleeting character of exchanges. Lyotard claims that '[t]he temporary contract is in practice supplanting permanent institutions in the professional, emotional, sexual, cultural, family and international domains, as well as in political affairs' (Lyotard 1984b: 66). While he acknowledges that the temporary contract is beneficial to the current system that operates via a performative principle, he argues for its potential in that it can also become the bearer of paralogical forms of legitimation, precisely because it is committed to a transient, localised and flexible structure.

How might the dematerialisation process, enabled through the information revolution, further enable a more equitable reconstruction of the social? One critic has attempted to extend Lyotard's claims arguing that an information society creates radical possibilities for the recognition of difference and plurality. Poster argues that:

> [t]o the extent that the mode of information constitutes a variety of multiple, dispersed, decentred, unstable subjects which contest the culture of identity, a new political terrain may be mapped in which the claims of the 'others' of postmodern politics must be placed in the forefront.
> (Poster 1990: 132–133)

Poster explicitly draws out the connections Lyotard makes between the reconstituting power of the information society and the potential for recasting social relations this allows for. However, it remains to be seen whether the radical reconstitution of social relations will lead to an empowering recognition of multiplicity, or a new form of oppression.

In the Introduction, I pointed out how a more abstract mode of subject constitution places the imperative on subjects to discipline themselves in order to construct their own identity. Lyotard does not consider how this process, while it escapes the disciplinary structure of modernity, leads to a novel situation whereby the subject implicates itself within a new disciplinary structure. Moreover, the radical openness enabled through the decentring possibilities of the informational realm does not necessarily lead to the recognition of others. The question remains as to how this particular mode of openness might work to construct an equitable form of postmodern politics. As Hinkson reminds us, we need to 'differentiate between levels of unfixity and among the modalities of openness – how unfixity can be carried by different levels of association with different structural implications' (Hinkson 1987: 152). Lyotard turns to a politics based on avant-garde aesthetics and the sublime. Whether this is an adequate response is an issue we must now consider.

Les Immatériaux

In 1985 Lyotard staged an exhibition titled 'Les Immatériaux' in Paris. The exhibition was specifically concerned with art and technology and some of Lyotard's attitudes towards it are worth considering, because they indicate the direction a postmodern politics, merging technological development with radical aesthetics, might take. The exhibition also tied together many of Lyotard's concerns, and allowed him to speak in a more direct manner about what he feels to be at stake in technological development. The exhibition was staged at the fifth floor of the Beaubourg Museum, which was transformed into a gigantic maze. Along the pathways of the maze were placed installations, electronic devices and cultural artefacts. The visitor wandered the maze hooked up to a set of headphones, which played a continual soundtrack of recited French cultural theory and modernist literature. The general aim of the exhibition was to evoke a feeling of uncertainty, resulting from the 'disappearance of a material world visible to the scientific gaze' (Jay 1992: 53). In many ways 'Les Immatériaux' forms a precursor for the issues that now concern Virtual Reality. The loss of objectivation, as a result of the advances in the techno-sciences, creates the potential to transform reality itself, especially the taken for granted sense of what it means to be 'human'. Lyotard summarises the social context upon which the exhibition reflects:

> The word 'human' . . . designates an ancient domain of knowledge and intervention which the techno-sciences now cut across and share; here they discover and elaborate 'immaterials' which are analogous (even if they are in general more complex) to those examined and detected in other fields. The human cortex is 'read' just like an electronic field . . . affectivity is 'acted' on like a complex chemical organisation.
> (Lyotard 1985: 49)

The context is one of dematerialisation and constitutive abstraction. The 'human' is 'seen' through, read like a code from a more abstract vantage point. The technological reading of the human body is not dissimilar from the flattening out of the organic body in the opening to *Libidinal Economy*. Certainly, according to the participants, the exhibition successfully evoked a sense of dematerialisation, effectively disorientating the subject's sensory body and the human sense of physical being in the world. For one commentator, 'the show was a phenomenologist's nightmare; everywhere one witnessed the replacement of lived functions of the body with artificial activities. One entered a world of simulation' (cited in Rajchman 1985: 115). But Lyotard would wish to distinguish himself from what he regards as Baudrillard's 'howling nostalgia' for a lost referent. The capacity of techno-science to transform every aspect of our lived sense of being ought not to provoke a Romantic revolt against technology. Lyotard has no use for the concept of alienation,[4] which underwrites such a revolt, regarding it as outmoded. On the contrary, he claims that the reconstituting ability of techno-science provides new possibilities for freedom. How can this be so?

Lyotard regards language as constitutive of our being, as determining our 'social bond'. Language paradoxically survives the electronic revolution. 'In essence', he explains, 'the new technologies concern language' (Lyotard in Rajchman 1985: 115). We can see how this is possible: the human body is 'read' as a code, human relations when conducted through electronic communication become more flexible and transient. Techno-science allows for a constant process of redefinition. This explains the presence of a large number of modernist icons at what was ostensibly a postmodernist exhibition. The dislocation of body, of a sense of locality, the impossibility of continual narrative are conditions rendered aesthetic in the modernist text, for instance in Beckett, whose works featured prominently. Lyotard's aim is to transpose the heterogeneous flux of identities registered in the modernist text into a radical social condition for postmodernity. The level of abstraction which allows for language to be constitutive of our being, that enables a constant flux of identities, is able, through technology, to be the dominant level through which we engage with the world. Rajchman summarises Lyotard's view of techno-science:

> as a modernist text in which we can continue the war of heterogeneous invention, once the province of advanced art – by other means. The danger . . . in techno-science is not an alienation of our supposedly natural identity; it is the 'totalitarian' possibility that there exists only *one* artificial identity that submits us to centralised control. It is the danger of a homogeneous unity of languages reflected in the informationist vision of controlled communication.
>
> (Rajchman 1985: 117)

Hence identity is not alienated, but opened up, subject to a continual process of reinvention. Technological development opens up the possibility of living life as a series of heterogeneous moments of interaction with fleeting others, as the subject constantly redefines itself. This is seen as advantageous by Lyotard, since it assists the dissolution of modern paradigms, in particular universalising narratives. Nevertheless, is this sufficient to provide us with a critical theory of technology? Does the polarity between the universal and the heterogeneous allow us to gain an adequate perspective on technology? For some critics, this is clearly not the case. Paul Crowther finds that 'Les Immatériaux' adopts an unquestioned acceptance of technological change: 'the questions the exhibition explicitly asks of the visitor (through the accompanying catalogue) concern only the nature of the changes taking place in contemporary life, and bypass the issue of their validity' (Crowther in Benjamin, A. 1992: 98). Indeed, if technology reconstructs the settings of social and cultural life, to the extent that the taken for granted and deeply held assumptions concerning our mode of being in the world no longer apply, can we speak of the 'advantageous' possibilities of the heterogeneous, in the manner that Lyotard intends? On this point, Hayles writes that Lyotard's apparent commitment to the 'denaturing of experience' leads to a contradiction. When Lyotard defined postmodernism as an 'incredulity towards metanarratives', Hayles writes:

> he had in mind such specific social narratives as the story of scientific progress and the rise of democratic education. But what are the essential components of narrative construction, if not language, context, time, and the human? The denaturing of experience . . . contributes a cultural metanarrative, and its peculiar property is to imply incredulity not just toward other metanarratives but toward narrative as a form of representation. It thus implies its own deconstruction.
>
> (Hayles 1990: 294)

This suggests that Lyotard's commitment to technological change, for its ability to break down universal narratives, may dissolve rather than fulfil the hopes he envisages for the postmodern. In other words, Lyotard's embrace of the techno-sciences, to the extent that they situate social and cultural practices upon a single abstract level, involve themselves in an ontological contradiction. However, we need to consider his aesthetic theory in more detail, in order to see if he can escape this apparent contradiction.

Aesthetics and politics – Duchamp and the sublime

Lyotard's writings on Duchamp provide a convenient way to register both his aesthetic position and how it might relate to the social. In *Duchamp's Trans/Formers*, Lyotard (1990) argues that Duchamp's works are exemplary because they foreground cognitive and perceptual insecurity. They refuse any assimilation into a grand narrative. Lyotard wants to transform this insecurity into a general ontological condition. If Duchamp's *Large Glass* creates a space of infinite optical regression, then Lyotard wants to collapse this aesthetic effect onto the social realm itself. The result would be, as Arthur Kroker notes 'a strange new world where subjectivity is lived as a relative projection at the hinge of deeply incommensurable realities' (Kroker 1992: 149). In this sense Duchamp can be seen as a precursor to the ontological relativism VR promises, and which technologies of extension such as the telephone and the computer already hint at.

If technology intervenes into human nature, if human being is to become to some extent 'immaterial', then for Lyotard the political choice is between 'machines'. Duchamp provides the model of 'a different kind of machine', a bachelor machine as opposed to industrial mechanics (ibid.). In opposition to the hegemonic goal of performativity, for communication without noise, Lyotard privileges Duchamp's art for its utter 'pointlessness'. Lyotard argues that Duchamp's 'machinery' calls into question the 'totalising and unifying machine' of contemporary techno-capitalism. The radical nature of Duchamp's machines are not confined to the aesthetic sphere however. Lyotard would wish to extend the principles of Duchamp's work to the social realm itself, arguing that it provides 'materials, tools and weapons for a politics of incommensurables' (Lyotard 1990: 18). In Chapter 4, I argued that attempts to locate two different machines within Futurist discourse (the open, productive and emancipatory machine against the closed and disciplinary machine), are limited in that they pay insufficient attention to the contradictory effects of an emancipatory project located within the more abstract framework enabled through the intervention of the machine. Can Lyotard escape this criticism?

Duchamp's work is important because of its constant search for new 'effects'. His machines are 'not enslaved-assertive but spontaneous affirmative and know no consequence'. The search for the new becomes the most desirable option in a model where 'there are only transformations, redistributions of energy. The *world* is a multiplicity of apparatuses that transform units of energy into one another. Duchamp the transformer does not want to repeat the same process' (Lyotard 1990: 36, emphasis added)

We have Lyotard's aesthetic neatly encapsulated in his work on Duchamp. The search for new effects is privileged, but, unlike the Futurists, this search is not goal driven. On the contrary, it is as marked by its very pointlessness. Anything less would be to recapitulate to the grand narrative of progress. Yet, if the aesthetic Duchamp so superbly encapsulates is to be transposed onto the social realm, then the question remains as to whether this aesthetic model can provide a viable model for social life? We may recall how the Futurists embraced a similar search for new effects that was, nevertheless 'grounded' by a contradictory ideology of aggressive subjective and national regeneration. If Lyotard is able to distance himself from the politics of the Futurists, can his model be sustained in the absence of any stabilising principle? Can a constant reformulation of our lived sense of being, which occurs, sustain an alternative model to the 'terror' of the modern. As Kroker so aptly puts it, Lyotard is 'haunted by the traditional problem of seeking a satisfactory synthetic principle for fractured experience' (Kroker 1992: 151). He turns to the sublime in order to try to resolve this dilemma.

It is clear that Lyotard's theory of the sublime is committed to a form of social relations that embraces the information mode. If information technology lifts us out of the prior settings which structured social life, then what is to prevent the rise of universality in this abstract realm? For Lyotard, the sublime maintains the incommensurability of different realms. What better theory for the age of the 'immaterial' than a rethinking of Kant's sublime, which described the 'shattering of belief' and the 'discovery of the "lack of reality" of reality'? The sublime can be contrasted to 'beauty' which depended on a universal consensus of taste and the capacity for reflective judgement. Given his hostility to universality, it is no wonder that Lyotard turns to the sublime as a means for acting 'without prior criteria'.

The sublime, according to Lyotard, is the place where Kant's attempt at bridging the faculties breaks down. In *Lessons on the Analytic of the Sublime*,

Lyotard traces the collapse of Kant's attempt to bridge the theoretical and the practical:

> [t]he relation of thinking to the object breaks down. In sublime feeling, nature no longer 'speaks' to thought through the 'coded writing' of its forms . . . Above and beyond the formal qualities that induced the quality of taste, thinking grasped by the sublime feeling is faced, 'in' nature, with quantities capable of suggesting a magnitude or a force that exceeds its power of representation.
> (Lyotard 1994: 52)

The social context of the information society with unbounded social relations and a conception of the self dispersed through circuits of exchange, allows Lyotard to return to a mode of thought where sheer magnitude exceeds representation. The action of the sublime shatters representation, hierarchy and normality, resisting all narratives which would seek to bridge the ideal and the real. The modern political conception which linked ideas with social practice (such as theorising the proleteriat) is rejected. Postmodern social relations have become impossible to represent. What takes the place of the modern paradigm is the aesthetic, where 'art is no longer in the service of cultural transformation; it *is* cultural transformation' (Readings 1991: 73). Lyotard insists that this sublime is not one that provokes a nostalgia for a 'real' reality which has been lost (a charge he levels at Baudrillard). Instead, this is a sublime that affirms the incommensurability of our projections in the world. If technology threatens the final instalment of the will to power, then the sublime counters this by affirming our ultimate impotence in the face of an unrepresentable reality. We can recall how Lyotard affirms the 'pointlessness' of Duchamp's machines. In the context of the information age, Lyotard is able to 'take the non-determined and limitless characteristics of the sublime, transform its universalising qualities into "local narratives" and project them onto the real' (Hinkson 1987: 139).[5] Hence, aesthetics are projected onto the social realm. What form will this projection take? In *Just Gaming*, Lyotard attempts to merge the desire for justice with the aesthetic desire for intensity and novelty. Ingram summarises his approach:

> what lends value to an action or work is its sublime violation of conventional taste, that is, its transcendence of the limits of representation and its shattering of established hierarchies of thought, disciplining boundaries and canonical norms into new configurations of discourse.
> (Ingram 1987: 289)

Can one really claim a place for justice in a world where the continual transcendence of limits constitutes the 'social' norm? Hinkson has argued that the previous experience of the sublime was available only to a minority – intellectuals and mystics who by reference to the cultural frames of their society were able to stand outside it (Hinkson 1987: 139–140). Lyotard's sublime, made possible through the information society, is a condition available to everyone. Hinkson argues that the form this will take is a heightening of the autonomous individual made possible through the 'image' society (Hinkson 1987: 140). This raises a serious question as to how a politics could possibly manifest itself through the sublime. Lyotard's sublime is an intensely personal experience, made possible by the way in which 'unstructured cultural realms seem to release the person into a medium of pure universality' (Hinkson 1987: 145). The conditions for such an experience necessarily arise through transcending the social. Lyotard hopes that the experience of the technological version of the sublime could dissolve the willing subject, which harnesses the logic of performativity, and so open a space for difference. Yet what allows this experience of the sublime is precisely what prevents any sort of social mediation of the experience. Axel Honneth comments on this dilemma, arguing that subjective freedom (which would enable the experience of the sublime as unconstrained and unrepresentable):

> increases only to the extent that [the subject] can leave the normative and cultural expectations of his time behind him while engaging in the innovative creation of possibilities for life, then the social lifeworld must be regarded primarily as a shackle for the individualising force of aesthetic self-invention.
>
> (Honneth 1992: 29)

The principle of incommensurability which comes through the aesthetic of the sublime, is one Lyotard hopes will clear a space for difference. Yet the manifestation of the sublime through the information society would lead only to a reified experience of difference. The conditions that allow for the postmodern sublime also dissolve the normative criteria through which difference could be read back into the social realm. Instead, the experience of the sublime is solipsistic. In this context, Meaghan Morris suspects the implicit return of a grand narrative in Lyotard's conception of the sublime. She argues that 'it is hard not to see why this ethic [of the sublime], revived for *art* would not again join the metanarrative of the speculative mind' (Morris 1985: 62).

Considering the role of the sublime, it becomes possible to say that, despite himself, Lyotard adopts a 'normative' notion of postmodernity. This is a postmodernity which guards against 'any and every hegemonic, closed or commensutory discourse, where there is always available sufficient counterpower to keep the game open' (Simpson 1995: 231). The argument being made here is that Lyotard's commitment to an aesthetic of incommensurability, embraces a form of constitutively abstract social relation which denies him the resources from whence any form of counterpower could arise.

Conclusion: reflections upon time

An essay in *The Inhuman*, entitled 'Time Today', is specifically concerned with how contemporary techno-science deals with the concept of time. Lyotard's concern is that the combination of technology and capital works to eradicate all forms of difference, including a differentiated concept of time. He points out that the universal language of the capitalist market is time. The capitalist emphasis is on optimising performance and this is achieved through the elimination of contingency. The possible complexity of the future is reduced to a linear model of a predictable future. Lyotard writes that:

> if one wants to control a process the best way of doing so is to subordinate the present to what is (still) called the future, since in these conditions the future will be completely predetermined and the present itself will cease opening onto an uncertain and contingent 'afterwards' . . . the present loses its privilege of being an ungraspable point from which, however, time should always distribute itself between the 'not yet' of the future and the no longer of the past.
>
> (Lyotard 1991: 65)

Lyotard argues that resistance can occur through keeping open the possibility of contingency and through preserving a differentiated notion of time. Yet, apart from calling for a 'temporal alterity' he does not specify how this process is to be achieved. He talks of how we must ensure 'the coming of the future in its unexpectedness' (Lyotard 1991: 77). Technology threatens this process, at least in its performative mode since the singularity of the present, the time of the event, is collapsed into a reified notion of time broken up into units of equivalence. Hence, the capitalist endeavour to 'save time' as if it were a unit of currency. Yet Lyotard seems to imply in *The Inhuman* that technology *itself* threatens the singularity of the present. In a significant passage marking a shift from the work on the postmodern, Lyotard asks:

> [w]hat does 'here' mean on the phone, on television, at the receiver of an electronic telescope. And the 'now'? Does not the 'tele-' element necessarily destroy presence, the 'here and now' of the forms and their 'carnal' reception? What is a place, a moment not anchored in the immediate 'passion' of what happens? Is a computer in any way here and now? Can anything *happen* with it? Can anything happen *to* it?
>
> (Lyotard 1991: 118)

It seems significant that Lyotard, who once talked about the advantages in the temporary contract, of the dissolution of the self in an abstracted network, and who remained committed to a framework of radical abstraction, is now concerned with relations structured by human *presence*. It is hard to see how one could constitute a differentiated notion of time through the abstraction of subjective and social relations. The quest for libidinal intensities might lead to a moment of *jouissance*, or an experience of the sublime, but the absence of ontological grounding would ensure that these moments could not be connected back to a personal history. As Hayles noted, the technological destruction of metanarrative threatens to undermine *all* forms of narrative. Consequently, the contextual frames which could 'value' time, would be emptied out. We have seen how the Futurist project was centrally concerned with the rejection of temporal difference and the embrace of a logic of pure flux and exchange. Habermas has identified the contradiction involved in such a commitment, noting that 'what is expressed in the value accorded the transistory and the ephemeral and in the celebration of dynamism is the longing for an immaculate and unchanging present . . . [a] "nostalgia for true presence"' (Habermass 1991: 344). Lyotard's position is contradictory in so far as his commitment to a technological framework allows for a more abstract experience of time that, in itself, could not provide the resources to constitute the temporal difference that he advocates in relation to capitalism. He has identified how such a system leads both to the domestication of time and to the further entrenchment of capitalism. Yet, it remains unclear how he might envisage an alternative, given his commitment to a radically abstract form of social life, where the *meaning* of time in relation to any act is constituted arbitrarily.

The theorisation of time is suggestive of thinking politically about technology without subordinating the political question to the issue as to whether or not universal narratives are dissolved. I have argued that Lyotard's radical politics are severely hampered by his embrace of an abstract social ontology. The alternative is not to reject this more abstract framework out of hand, as Virilio, who I examine in the next chapter, tends to do. Rather, we need both to go beyond a position that simply celebrates or denounces the liberation of the individual from all prior modes of being and to consider whether certain historically and culturally embedded needs can be met within such a framework. What is required is to theorise reflexively the deeply embedded human desires for transcendence and to work to construct forms of sociality that might accommodate them without being framed by commodity fetishism or cultural despair.

6 Paul Virilio
Overcoming inertia?

> Will Earth soon become humanity's phantom limb?
> (Virilio 1995: 127)

Paul Virilio has remarked that 'we should, after two centuries of positivism, progressivism and idealism of techno-science, come to critique the negative aspect' (Virilio in Madsen 1995: 80). Much of his work has been devoted to highlighting the negative aspects of technological development and change, writing with a certain apocalyptic sense of what an extrapolated technological future might hold. Virilio's work has become increasingly influential, going well beyond its initially favourable reception in the avant-garde circles of art and social theory. Yet, there has been relatively little discussion as to the overall value of Virilio's critique of technology. Certainly, there is much to be said for his sustained opposition to the increasing colonisation of life by technology. At a time when new technologies are the subject of an enormous amount of uncritical hype, when even 'critical' intellectuals seem content to revel in the play of technologically-mediated simulations, Virilio's is almost a voice in the wilderness. But despite his consistent critique, and pointed observations concerning the 'negative aspect' of technology, I want to argue that Virilio's work remains limited in the degree to which it can contribute to a critical or ethical engagement with technology.

More than any other theorist discussed in this book (except perhaps Heidegger), Virilio has focused on the role of technology as a reconstituting agent, in relation to embedded social and cultural meanings. In this sense, his work is vital in that he sketches out the ground on which we can assess the impact of technological change. However, his work never goes beyond this point. So while Virilio is valuable for one part of my argument (critique), it is severely limited in terms of outlining the ground for a more reflexive theory of technology. Indeed, this limitation at times makes Virilio partially complicit with the trends and ideologies he opposes. Virilio's aphoristic brilliance allows his theoretical insights to reveal themselves spectacularly, like the technological 'accidents' which briefly counter the prevailing technological *telos*. As with the nature of accidents however, Virilio's rapid fire missives fade almost as quickly as they appear.

This book has argued for a need to conceive of the social as an intersection of levels, each more or less abstract. It is the existence and often contradictory relationship between these levels that is downplayed in Virilio, remaining only implicitly available through his category of 'disappearance'. Hence, he develops only a one-dimensional critique, one that cannot determine how technology might be deployed critically, or show how effective forms of resistance might be allowed to unfold beyond a mere strategy of invocation. I hope to show this through an extended analysis of Virilio's work, arguing that his pessimism and lack of alternative is a *structural* problem rather than a question of individual inflection. This structural difficulty can be further highlighted through an analysis of the writing of McKenzie Wark, who attempts to use Virilio's work to gesture towards an alternative conception of technology lacking in Virilio himself. Ultimately, I want to say that both Virilio's overwhelmingly negative analysis and Wark's optimism are insufficient because they fail to go beyond the construction of a one-dimensional ontology that overstates the extent of technological reconstitution. Such an ontology can only move between a technologically determined nihilism (Virilio) and the celebration of heterogeneous fragmentation (Wark).

Virilio's critique of technology can be loosely divided into three approaches. First, as the opening quotation has shown, he is determined to reveal technology as an ideology, bound within the limited frameworks of 'positivism, progressivism and idealism'. In addition, Virilio links technological development with a distinctly militarist ideology, arguing that our modes of perception in daily life are altered by technologies developed for military purposes. The result of this is to reorganise our perceptory functions within a militarist framework, which blurs the distinction between war and peace. Thus, technology constructs a materialised ideology that constructs a proto-military representation of the world. Kafka's claim that 'cinema involves putting the eye into uniform' is cited by Virilio (2000: 29) to illustrate this point. In *The Art of the Motor* (1995), Virilio observes a distinctly metaphysical construction of new technologies in what he calls 'techno-fundamentalism', a substitution of technology for religion at the centre of metaphysical thinking. Second, Virilio regards technology as a means of 'colonisation': of the categories of time and space; of habitats such as the home, the city, the environment; of the senses, particularly vision; and finally of the body and the human subject. Virilio can be referred to as an ecologist, but one whose conception of threatened habitat extends well beyond the physical realm, to ontological categories such as time and space. Finally, Virilio wishes to highlight what might be called the *point of reversal* where technology, conceived of as a tool, produces the opposite effect of that it was originally designed to facilitate. Hence, to take the most well known of Virilio's claims, technological speed – our ability to engage with the contents of the world more quickly through faster transportation or information processing – flips over into a state of 'inertia' – where we are immobilised both physically (at a computer desk) and ontologically (via a dispersal of constitutive frameworks). This example can be understood through the terms of the levels of constitutive abstraction argument made in the book: where 'speed' across disembodied, abstract hyperspace results in intertia at the level of the face-to-face. It is Virilio's failure to come to terms with the contradictory relation between these levels that leads to the one-dimensionality of his critique.

At times, Virilio connects these process of colonisation and reversal to some form of social process, such as late capitalism, or the rise of the military-industrial complex. But at other times, technology seems to be regarded as an autonomous and determining force, and as such he veers towards a spectacularly updated version of Heideggerrian pessimism. We shall evaluate these approaches after we examine Virilio's work a little more closely.

Technology as an ideology

If philosophers have for some time pronounced the death of God and metaphysics, this move, according to Virilio, has not led to the evacuation of a transcendental centre. On the contrary, it is technology which comes to occupy the centre of metaphysics, at the expense of humankind and nature. Virilio claims:

> [f]aced with the demands of an increasingly artificial terrestrial environment and the disastrous consequences of a criminal level of pollution ... will we see a new type of FUNDAMENTALISM emerge, one no longer associated with the trust in God of traditional beliefs, but with the worship, the 'technocult' of a perverted science?
>
> (Virilio 1995: 118)

Virilio argues that the latest and most pervasive form of technology as ideology is techno-fundamentalism. Such fundamentalism is, for Virilio, merely a furtherance of the scientific will to power, but one that begins to become the dominant frame through which our existence is measured. Such fundamentalism, far from being necessary to our survival, will work towards obliterating all that is left of the human 'species'. It will 'attack what is alive' (ibid.), eliminating any 'natural vitality'. Virilio argues that, rather than furthering human capacity through technological extension, techno-fundamentalism will erode our embedded constitutive frameworks, so that the 'human' will become subordinated to the technological, rather than the reverse. As evidence for this tendency towards technological fundamentalism, he cites developments in the sciences (genetic engineering, organ transplants) in cybernetics and in aesthetics, focusing in particular on the performance artist Stelarc.

In the essay 'From Superman to Hyperactive Man' (1995), Virilio examines the colonisation of the human body under the processes of technological speed. In particular he examines the ideologies contained in the performance art of

Stelarc, which he compares to the Futurists. Stelarc's aim is to explore the aesthetic and philosophical possibilities of a technological reconstitution of the body, with the intention of transcending the limitations of our present human condition, so as to enter the post-evolutionary era of the cyborg. Virilio quotes Stelarc:

> I have begun in the course of my performances to ask questions on *the design of the human body* and the more I work, the more I believe that the body is becoming obsolete ... the ultimate limit of the body is the physiological limit, our weak organic capacities ... Technologies have always been outside of the human body, but now technology no longer explodes far from the body. This is perhaps the most important event in our history: it is no longer necessary to send technologies towards other planets but to make them land on our body.
>
> (Stelarc in Masden 1995: 80)

Stelarc aims for a technological transcendence of the limits of the human body, but, for Virilio, he represents 'a kind of suicide of the body' (ibid.). Virilio critiques the hubristic focus on the increased range of possibilities for the technologised subject. Instead of an increase in subjective possibility, Virilio argues that Stelarc is caught up in a dangerously progressivist ideology. He claims that:

> [w]hen Stelarc says 'I want a stimulator that will give me a more powerful heart. I want more powerful eyes which will be like telescopes ... I want the most extraordinary sex so that I can make love to all women' ... this is the atomic bomb again. It is the end of humanity in a totalitarian vision, it is a fascist phenomenon.
>
> (ibid.)

All of Virilio's work is an attempt to critique the logic of domination inherent in the technological will to power. Where many see technology extending human capacities, Virilio sees exploitation and habitual destruction. More importantly, he sees the possible return of a fascist logic bound up in the Futurist aesthetic, which celebrated the coupling of the human and the technological in the war machine. In this context, Stelarc aestheticises the extension of the war machine into the body itself. Yet Stelarc is, according to Virilio, merely a victim of the current ideology of technology, 'a victim of the situation' (ibid.) This 'situation', that is, the aim for technological transcendence over physical and biological limits, is similarly manifest in the utopian discourses that surround cyberspace, virtual reality and the science laboratory at MIT (see Chapter 7). Stelarc's aim for technological transcendence is merely an aesthetic representation of this broader phenomenon.

Stelarc's technological performances take on a darker reality according to Virilio in the case of the remote-control suicide machine created by Philip

Nitschke. This technologically enabled form of euthanasia takes away the dimension of human agency. For Virilio, it 'wipes away the patient's guilt, together with the scientist's responsibility' (Virilio 2000: 5). Such technological domination also reveals itself in the case of the world chess champion Kasparov 'playing a game against a computer specially designed to defeat him' (Virilio 2000: 5). Both examples reveal a kind of euthanasia of human subjectivity, the first literally, the second metaphorically.

For Virilio, technological expansion is inextricably linked with colonisation, not just of exterior or outer space but of our inner selves as well. Echoing both Heidegger's and the Frankfurt School's equation of the domination of outer and inner nature, Virilio observes that '[w]e have never, in fact, dominated geophysical expanse without controlling . . . the microphysical core of the subject being'. He lists many examples of such control, beginning with the domestication of other species, through the disciplining of the soldier, to the 'necessity' for athletes to take anabolic steroids. Stelarc is thus merely the latest and most literalised example of the intertwining of inner and outer control, masked by the ideology of expanded freedom made possible through technological mediation. New technologies such as cyberspace and genetic engineering, rekindle the Futurist dream of transcendence over prevailing conditions. However, for Virilio, there is one important difference.

If the Futurists embraced a particular *aesthetic* ideology, Virilio argues that this ideology has become increasingly centralised, and more importantly *materialised*, through contemporary technology. He writes that:

> the 'new machine' has certainly materialised the cutting loose initiated by Futurism, Cubism or Surrealism, but now it is less a question of dissociating objective appearances from reality, from the artist's subjective interpretation, than of shattering *man's unity of perception* and of producing, this time AUTOMATICALLY, the persistence of a disturbance in self-perception that will have lasting effects on man's rapport with the real.
> (Virilio 1995: 146–147)

The shift from an earlier Futurist ideology, to the more pervasive ideology that surrounds contemporary conditions entails a *qualitative*, as well as quantitative, shift in our relation to technology. While the Futurists' technological *imaginary* saw no bounds to the extent of technological mediation, they were constrained historically and materially, as to the degree of technological transcendence possible. While technologies such as the motor car, the cinema, and modern weaponry allowed a modification of our modes of engagement with the world, such transcendence was very much defined through reference to a relatively stable normative framework. One couldn't live permanently in a motor car (a privileged site of Futurist transcendence), for instance, whereas today it is almost possible, and considered desirable by some, to spend one's whole time in cyberspace, or to be 'rewired' though genetic engineering.[1] In other words, it is increasingly possible to live a mode of existence 'cut loose' from the prior social and cultural frameworks that grounded traditional and modern forms of individual and collective meaning. Virilio argues that the technological forms of transcendence that are available, and which partially structure contemporary life, create a totalising phenomenon, in that no dialectic is possible between aesthetic transcendence and more grounded modes of being in the world. The essential point for Virilio, is that this contemporary form of technological transcendence happens, 'automatically' as he puts it. The result of this ideology of transcendence is a crisis of meaning and a lack of ethical and aesthetic references to live by. If the Futurists celebrated technological war in order to reassert the nation-state, Virilio argues that today it is barely possible even to conceive of a bounded and definable notion of space, through which we could ontologically comprehend the idea of the nation.

One of Virilio's most important observations about ideology, however, goes beyond a critique of idealist progressivism. For Virilio, ideology is not confined to an abstract pattern of belief. Instead, he argues that specific ideologies, such as technocratic rationality and militarism become *materialised* through technology. Technology works to reconstitute our sensory perceptions so that such ideologies are naturalised, in a manner that goes beyond even Althusser's most pessimistic conclusions about the way ideology is practically lived out (see Althusser 1971). This observation, concerning the manner in which technological mediation allows for the materialisation of specific ideologies, is at its clearest, in Virilio's work, when he discusses the intertwinement between the military and the development and implementation of new technology. He comments on how the 'virtualisation' of everyday life, that is, the increasingly technological mediation of the world via media images and communication systems, is made possible through the cohabitation of the military and the technological, not only because military research can be carried over into the civilian sphere, but also because the technological reconstitution of the real constructs the social realm as a quasi-militarised space. Virilio writes that:

> the tangle of [communication] networks blackening your map is only the triumph of the military population, the administration of a territory set up for the conductability of war . . . this is what the doctrine of security is founded on: the saturation of time and space by speed, making daily life the theatre of operations, the ultimate scene of strategic foresight.
> (Virilio 1990: 92)

The militarist ideology, the collusion of technological development with militarised terror, extends across the social realm as Virilio traces the passage from 'wartime to the war of peacetime' (Virilio and Lotringer 1983: 142). The reconstituting ability of the war machine has disseminated across the social field itself. This is what Virilio means when he talks of society as in a perpetual state of war, 'all of us are already civilian soldiers, we don't recognise the militarised part of [our] identity'. This aspect of our identity is constructed through technologically reconstituted modes of perception and engagement, modes first developed in the military.

In *War and Cinema* (1989a), Virilio makes the now familiar claim that, in the twentieth century, seeing is the dominant means of gaining knowledge. He goes on to argue that, if new technologies modify perception, they correspondingly help to construct a new 'real'. For Virilio, the new modes of perception are inextricably linked to military developments in technology. Ideology in this sense is materialised: the way we engage with the world is framed by technologies of perception, particularly vision. In *War and Cinema*, Virilio theorises war as that which is capable of 'scoring territorial, economic or other cultural victories as in appropriating the "immateriality of perceptual fields"' (Virilio 1989a: 7). It is the capacity of military technology to appropriate our perceptual fields that concerns Virilio. He makes a convincing case for the collapse of militarist perceptual techniques into civilian life, but we can also see the opposite, as we know from 'watching' the Gulf War, where the war was constructed through western media as 'cinema'.

Ultimately, for Virilio, the progressivist and idealist understanding of technology, which regards it as a useful and beneficial instrument of progress masks a more baleful reality where the technological reconstitution of human life leads to inertia, powerlessness, a loss of a meaningful conception of what is real, and a militarised form of subjective engagement within civilian life. As we have already seen, Virilio's work extends beyond a critique of 'ideology', in the traditional sense, and examines the way our modes of relation with the world have become increasingly mediated or, in Virilio's opinion; *colonised* by technology. Broadly, Virilio's work has focused on the following areas: time, space, movement, social relations and the body. Virilio argues that technology's capacity to transcend limits has led to the increasing colonisation of the earth and of the human species. Much of his work is devoted to describing this process, largely through concentrating on the technological reconstitution of time. While Lyotard's work, *The Inhuman*, has also examined the colonisation of time and its relation to the capitalist structure of exploitation, Virilio is more concerned with the way technological time can empty out the ontological frames of meaning for the human subject. What is interesting here is how Virilio grapples with this process through the concept of 'disappearance'.

Time

Time is the crucial category for Virilio. The relationship between technology and time affects not only our sense of temporality, but also our relations with ourselves, our bodies, others and the earth. One of Virilio's central concerns is how time is reconstituted, through technology, into what he calls 'speed'. Speed describes the technological reconstitution of our temporal mode of engagement with the world. The creation of faster types of transport, greater flows of information and images by technology has led, according to Virilio, to a qualitative shift in the way we act and perceive the world. Speed describes a flattening out of temporality into a one-dimensional framework. The value of something is no longer judged according to its specific content, but according to its form, since things are only understood in terms of how quickly they can be made to function or how quickly they can be accessed. This mutation affects us in several ways, according to Virilio: it affects our sense of perspective; it robs us of any sense of specific place; and it flattens out our lived sense of being, so that the human subject exists merely as a kind of 'motor' that can change speeds in a quantitative but not a qualitative sense. We are framed by technological speed, even when we are not using any particular technology, in that instantaneous communication, the ubiquity of interactivity and the interface, the erasure of phenomenological distance, have left us with nothing from which to re-establish an alternative framework, through which we could engage differently with the world.

Virilio recognises the ontological importance of a certain kind of temporality, in that rhythmic time 'structured the life of former societies' (Virilio 1991: 82). As he puts it, '[b]asically, time is lived – physiologically, sociologically and politically – to the extent that it is interrupted' (ibid.). Modern technology is precisely that which alters the phenomenon of lived time. Time is rendered more abstractly, it gives us new phenomena such as the 'third window' the 'electronic day' and the 'false day'.

The reign of technological speed subjects us to the new 'regimes of temporality that issue forth from advanced technologies'. Hence the phenomenon of the *electronic day*, which restructures time according to an artificial technologically-rationalised sequence. The connection of time to natural or phenomenological measure is severed. According to Virilio, 'the new technological time has no relation to any calendar of events or to any collective memory' (Virilio 1991: 15). One example is the lack of differentiation between night and day. Virilio remarks that technology 'kills the night'. Historically and culturally embedded rhythms, structured through temporal differentiation are threatened by technologies that can allow subjective activity to unfold regardless of when or where they are taking place. Ultimately, 'the exhaustion of natural relief and of temporal distances creates a telescoping of any localisation, of any position ... The instantaneousness of ubiquity results in the atopia of a single interface' (Virilio 1991: 17). Virilio seems to be saying that when everything is instantaneously available, relations of equivalence structure the relations between things, so that the transcendence of temporal constraint leads to a generalised sameness.

Virilio is a radical ecologist in that he argues for an ecosystem that extends well beyond the physical environment. He argues that we need to preserve an ecosystem, not just of the physical environment, but also of space and time. Information technologies in particular, through their ability to reach across time and space, create a new type of pollution; what Virilio calls the 'pollution of distances'. He writes that: 'Today what is threatened is not only the planet, the water, the air ... which concerns the ecology, but space itself.' How is space threatened?

Space

For Virilio, technological time, that is time lived as a one-dimensional experience of 'speed', empties out the ontological category of space. Virilio historicises the relation between technology and our understanding of spatial categories. He describes a shift, from the ordered and homogeneous space, the space of classical Greek geometry, to a technologised space that is accidental and heterogeneous. The destruction of urban space is less due to any physical factor than to the reconstitution of space into technological time, where 'the screen abruptly became the city square, the crossroads of all mass media.' Space is colonised by technological speed, particularly through new technologies that allow the rapid movement of information and the easy traverse of distances. As Conley remarks, 'space collapses into time, for when instantaneous communication and supersonic travel are commonplace – all cities exist in the same place – in time' (Conley 1993: 87). While space still exists, the meaning of space as a category that can frame our mode of engagement with the world, 'disappears'.

Virilio extends this point to discuss how technology has altered our relation to culturally specific spaces, such as the city. He writes that:

> [i]nstead of operating in the space of a constructed social fabric, the intersecting and connecting grid of highway and service systems now occurs in the sequences of an imperceptible organisation of time in which the man/machine interface replaces the facades of building as the surface of property allotments.
>
> (Virilio 1991: 63)

Technology, especially communications technology, has altered the meaning of the city. Once a centre of social and mercantile exchange, the city as a meaningful *site* has been undermined by technologies that allow subjective actions to be carried out regardless of their specific location. The circulation of electronic capital and information exceeds the constraints of physical space and location and, in the case of the city, alters the cultural meaning and subjective experience of the city as place. Scott Lash notes how for Virilio, the chief virtue of the city is the fact that it is enclosed. For Lash 'Virilio's notion of the "good life" . . . is bound up with this metaphysics of enclosure and interiority' (Lash 1999: 286). Technology erodes the interiority of both the city and the individual enclosed within it.

For Virilio, 'the city is now composed of a "synthetic space-time" that *simulates* the lost geophysical urban spaces of human habitation and circulation' (Bukatman 1993: 126). The exchange of goods and information still occurs, but is not tied to any specific location. In other words, 'people occupy transportation and transmission time instead of inhabiting space'. While this may allow more freedom and possibility, in some senses, Virilio argues that the reconstitution of exchanges into a more abstract space-time which undermines the significance of something like the city has oppressive possibilities. The delocalisation of actions might lead to a 'liberation' from determined spaces, but it creates a new space that may, in fact, constrict subjective freedom. On this possibility Virilio argues that:

> the physical geography of France has completely disappeared under the inextricable tangle of the different media systems; that *not only does delocalisation occupy more territory than does localisation, but it occupies it in a more totalitarian fashion.*
>
> (Virilio 1990: 92)

He sees a danger in the creation of a new, technologised, abstract space, in that it severs our relation to prior spaces. It is not that such spaces no longer exist, but that there is a danger of information technologies undermining the social and cultural frameworks which rendered the experience of 'place' meaningful. This is the 'totalitarian' possibility, that expanded freedom in an abstract space only allows for freedoms within a single plane, the virtual space of information technologies. If Lyotard looks to information technologies in order to escape totalitarianism, Virilio locates it precisely within these technologies.

This book has argued that more abstract constitutive levels draw upon and reconstitute ways of being and acting within less abstract levels, and that this is a contradictory process. Moreover, I have suggested that theorising these contradictions may enable a 'third way' beyond Lyotard's empty radicalism and Virilio's pessimism. While Lyotard embraces this reconstitutive process, and attempts to situate a radical politics within this more abstract level, Virilio despairs at whether we can even make meaningful choices within this abstract level. Neither theorist concerns himself with the *contradictory* nature of this reconstituting process.

Virilio's term to describe the meaningless relations that take place at this abstract level is *inertia*. It resembles a postmodern version of Heidegger's 'functioning' and, like Heidegger, does not sufficiently grapple with dilemmas surrounding technological change.

Movement (and inertia)

Technology as 'speed' ultimately leads to a process of ontological levelling that culminates in inertia. Humanity begins to share the fate of the geographical poles which remain still while the rest of the earth rotates – this polar inertia becoming a universalised condition. For Virilio, technological progress infinitely promotes speed, as the subject passes from a perceived freedom of movement to a tyranny of movement. As Virilio points out 'the field of freedom shrinks with speed. And freedom needs a field. When there is no more field our lives will be a terminal, a machine with doors that open and close' (Virilio 1993: 54). Subjective experience is now framed by technological processes in a manner previously unimaginable and, for Virilio, the reconstitution and simulation of subjective engagements with the world works to speed up the *process* of engagement, but only by immobilising the subject as a necessary condition of this speed.

The inertia Virilio talks about takes several forms. First, there is a perceptual and sensory inertia, where the subject's sensory apparatus, having been deprived of its culturally embedded contextual framework, by having to operate at a more abstract technological level, is unable to assimilate the information it is given. Accordingly, Virilio discusses the 'pathology of movement' that arises from modes of prosthetic travel. He notes that '[f]rom the elimination of the physical effort of walking to the motor loss induced by the first fast transport, we have finally achieved states bordering on sensory deprivation' (Virilio 1995: 85). This is one example of how, for Virilio, technologies which are designed to extend the capacities of the human sensorium, can, through transcending the context in which the subject normally processes information, actually restrict the subjective ability to interact within a given environment. Virilio compares the handicapped subject with the cyborg subject of techologised warfare, in order to highlight the disciplinarian framework of full-blown technological mediation. He claims that the modern warrior has been transformed from the very essence of physicality to a technologically-dominated subject which remains 'tied to his machine, imprisoned in the closed circuits of electronics the war pilot is no more than a motor-handicapped person analogous to the hallucinatory states of primitive warfare' (Virilio 1986: 85). The technological extension of the senses depends upon a certain *pacification* of other aspects of embodiment, as we shall see in the chapter on Virtual Reality. In this sense, Virilio's initially extreme comparison does reflect the confluence between technological extension and technological domination.

The second form of inertia is temporal inertia. Virilio discusses how the effects of technological speed can result in the colonisation of the future. Sounding very much like Baudrillard, he claims that our technological ability to transcend time robs us of any sense of perspective or anticipation, so diminishing our ability to imagine any alternative to the present. For Virilio, 'speeding is precisely the elimination of expectation and duration'. Our technological ability to speed through time alters our relation to time, particularly the subjective sense of lived or embodied time. Virilio explains: '[f]rom now on everything will happen without our even moving, without us even having to set out' (Virilio 1989b: 112).

This technological relation to time constructs time not as something to be lived (or moved) through, as a medium and ground in which actions meaningfully unfold, but instead as detached from action, a barrier to subjective goals and thus something to be overcome. Hence, temporality is reduced and understood only through the framework of technological speed. This reconstitution affects our subjective relation to time. For Virilio, we will come to experience time prosthetically: 'it is our common destiny to become film' (ibid.). Such prosthetic temporality will further erode our ability to construct a relation to the real, or even to the past, for as Virilio warns 'if time is history, speed is only its hallucination' (Virilio 1996: 40).

Virilio's description of the relationship between speed and inertia serves as a useful corrective to more naive appraisals of the revolutionary capacities of information technologies. Nevertheless, he fails to explore in more depth the contradictory nature of 'speed'. It is possible to think about 'speed' more productively in terms of the levels of abstraction argument. For instance, the excitement of 'speed' depends upon the continuing sense of embodied limitation. In this sense it is very similar to the excitement of 'cybersex'.[2] An approach that reflexively theorised these levels-in-contradiction, whereby we recognised that the benefits of technological speed are best achieved by allowing more concrete forms of interaction to limit the degree to which disembodied exchanges become the dominant mode of engaging the world, presents a more constructive approach than Virilio is able to furnish.

Social relations

The sensory immediacy which new technologies permit as we extend our sense of community across the globe, corresponds to an increasingly abstracted notion of the 'others' who form this community. The other brought nearer through technology is a fundamentally different entity. We no longer deal with them in the same manner we would at a face-to-face level. Virilio charts the changing nature of the social relations new communications technologies construct, where 'the presence of absence, whose unlimited abundance is revealed by the new means of instantaneous inter-communication, such as telematics, citizenband radio, Walkman and video technologies' (Virilio 1991: 94).

However, this presence of absence, which allows for things, sounds, and others to appear also affects the manner through which we engage with them. For Virilio, the excess of sociality made possible through technologies of extension disperses the capacity for meaningful social relations. He claims that 'we will never be neighbours in televisual proximity' (Virilio 1991: 84). Virilio is referring to a shift in the framework of sociality, where we increasingly engage with the mediated presence of intangible others across space and time. He argues that this shift alters the way we relate to more concrete others. In other words, 'when things "far" are brought into immediate proximity, those that are proportionately "near" such as friends, kin, neighbours – turn what is proximate – family, work, or neighbourhood – into a foreign, if not inimical space' (see Virilio in Conley 1993: 10–11). For Virilio, the tangible 'presence' of another becomes an alien, threatening, presence as technology causes social integration to be carried out at a more abstract level.

Yet it is here that we can begin to question the terms of Virilio's analysis. While he recognises a shift in the way social relations are carried out, he can only express this as an oxymoron, 'the presence of absence', and relate it to the activity of speed. By doing this, he theoretically constitutes social relations within a single abstract plane. It is simply not the case that 'we will never be neighbours' through the use of tele-communications. Rather, the social bond is constituted more abstractly, but we relate to these intangible others *as if* we could carry on a face-to-face discussion. In ignoring the contradictory manner through which abstracted modes of integration are constituted, Virilio misses out on a crucial critical perspective, as we shall see below.

The body

Ultimately, this process of technological mediation renders intangible not only the other, but also our relation to ourselves, particularly to our bodies. Virilio writes:

> In days gone by, *being present meant being close*, being physically close to the other in face-to-face, viz a viz proximity. This made dialogue possible through the carrying of the voice and eye contact. But with the advent of *media proximity* . . . we can not only act at a distance, but even teleact at a distance, see, hear, speak, touch and even smell at a distance – then the unheard possibility arises of a sudden splitting of the subject's personality. This will not leave the subject's 'body image' – the individual's SELF-PERCEPTION – intact for long . . .
>
> (Virilio 1995: 106)

It is at this point, where technology facilitates a novel form of self-alienation, that we can turn to Virilio's discussion of the next 'revolution' that of organ transplantation. In *The Art of the Motor* (1995), Virilio focuses on how new technologies of transplantation and implantation have allowed for the further colonisation of the human body. For Virilio, there is a familiar pattern: the body is attacked in a similar manner to space and time, in other words, we have technology reconstituting a heterogeneous field of sensation, perception and experience and flattening it out into a one-dimensional ontology. In this case the rich and varied experience of human embodiment is harnessed to the imperatives of technocratic rationality. Organ transplants and genetic engineering are designed to maximise the efficiency of the human body as instrument. Virilio uses the metaphor of the body as a motor to highlight its reconstitution through technological speed. He writes: '[n]ow we are going to treat the human being as if he were a motor, a machine to accelerate constantly' (Virilio 1995: 103). The metaphor of the motor captures an appropriate sense of dehumanisation, but also conveys the reconstitution of the body, as a site of heterogeneous forces into an objectified field where change can only be measured by quantitative difference, that is, a change in speed.

The human body, in particular the inner body, represents perhaps the final site of technological colonisation. It comes to be regarded as a kind of motor, harnessed to the technological, the final completion of the human–machine interface, begun in the information revolution, where the subject increasingly interacted with media technologies. Virilio remarks:

> [b]ringing the body and its vital energy up to speed with the age of instant teletechnology means simultaneously abolishing the distinction between *internal* and *external* while promoting a final type of centrality, or more exactly, hypercentrality – *that of time* . . .
> (Virilio 1995: 119)

As with the field of action created by information technologies, this is not a natural or lived time of the body, but rather an abstract technological time. If technology-as-speed colonised outer space, shrinking space, destroying a sense of specific place, invading and overcoming the city, and rendering obsolete architecture as a constitutive frame, then a similar process will overcome the constitutive meanings of embodiment of our 'inner space', as the 'natural rhythms of the body are replaced by technology and harnessed to its hyper-rational imperatives'. Virilio is appalled by the prospects of a future where 'the technosphere prevails over the biosphere' (Virilio 1995: 115). He argues that our inner bodies are threatened with a similar process of constitutive dispersal to our external perceptual fields. As an example of this dispersal, Virilio discusses the phenomenon of 'intraorganic dislocation', the sense of alienated embodiment felt by subjects who have undergone organ transplantation (Virilio 1995: 107). He discusses the experience of a transplant patient who feared that the grafted organ was 'falling' and who remained frightened that the new organ was going to unattach itself. Virilio argues that the body has a vital centre of gravity with its own sense of inner referentiality, a centre that is now coming under threat.

The distinction between 'I am a body' and 'I have a body' has already been problematised in contemporary culture since the organic body has been reconstituted (both theoretically and technologically) as a series of zones of signification, or as we saw with Lyotard, has been characterised as containing multiple zones of libidinal intensity. The tendency within postmodernity is to reconstitute the body as an abstract series of manipulable parts. The transplantation revolution promises to abstract our sense of lived embodiment even further as our 'insides' come under the sway of technological reconstitution. 'Sooner or later', Virilio claims, 'intimate perception of one's gravicentric mass will lose all concrete evidence, and the classic distinction between "inside" and "outside" will go out the window with it' (Virilio 1995: 106).

In providing a critique of the way technology has colonised recognisable areas of 'habitat' such as time, space and the body, Virilio seems to be arguing for the ontological necessity of scale, proportion and horizon. It would seem that, as a self-styled 'art critic of technology' (Madsen 1995: 80), he wishes to oppose the making over of the world into one giant canvas of abstract expressionism. Hence, he warns that: 'with the interfacing of computer terminals and video monitors, distinctions of *here* and *there* no longer mean any thing'

(Virilio 1991: 13). The result of the removal of embedded referentiality is a series of reversals.

Virilio argues that the increasingly technological mediation of our worldly engagement inevitably tends towards what he calls *constitutive dispersal*. Such constitutive dispersal describes a world without measure, 'an open system in which no one can find any perceptible, objective limits' (Virilio 1991: 72). Virilio traces a characteristic 'point of reversal', where the faculties technology is originally designed to extend actually begin to atrophy. Throughout his work, Virilio has focused on the point where results become the opposite of their original intention. Hence movement (temporal and phenomenological) becomes inertia, extended social relations culminate in social alienation, as we become encased in tele-visual proximity, electronic data and information produce tele-amnesia, technologically-enhanced bodies resemble 'handicapped' machines, while heated forms of human conflict (war) turn into a cold preparation for perpetual war and a hyper-militarised civilian force.

The conclusions Virilio consistently points to – colonisation, disappearance and reversal – are not as inevitable as his more technologically determinist passages seem to suggest. Unfortunately, Virilio provides no coherent theory of how this future might be avoided. The next section will examine the problems in his work and will suggest that a theory of technology, based around the theme of constitutive levels, is more open to the construction of a reflexive critique of technology than one which focuses too exclusively on ideology, colonisation, disappearance and reversal.

Resisting technology?

Virilio's critique of technology is important in several ways. He is right to oppose progressivist and idealist constructions of technology, especially when technological utopianism seems to be on the rise with the advent of digital communities and medical and genetic 'advances'. Moreover, he has effectively made a strong case for how the technological reconstitution of the senses brings with it a pervasive disciplinary framework, one that is both quasi-militarised and empties out the ontological frames that formerly grounded social and cultural meaning. Yet, while Virilio's work is important for the sheer range and insightful nature of this critique, his overall strategy remains problematic. My critique of Virilio is divided into three sections: first, the overstated nature of his claims; second, the limited value of his strategy of denunciation and of the way it remains trapped within a metaphysics of heroic subjectivity; finally problems of a terminology which limits our ability to move beyond the one-dimensionality he wishes to critique.

The politics of overstatement

There is no doubt that Virilio writes at times with a certain apocalyptic fervour. While the reconstitution of our habitat, subjective framework and body by technology is doubtless occurring, this is not so comprehensive a process as Virilio would have us believe. No doubt Virilio's is a deliberate strategy, a kind of cathartic nihilism intended to prevent the actual outcome of what he describes. As Conley points out, much of Virilio's work 'mimes the very system he criticises by projecting his way of apprehending the present into futuristic visions' (Conley 1993: 86). Yet this strategy can only be limited and has its own dangers. In particular, there is the danger of succumbing to technological *determinism*. While Virilio might be horrified by the technologised future, as he imagines it, his *descriptions* are in essence no different form those who succumb to a progressivist ideology of technology. His discussion of Stelarc is illustrative of this point: while he repeatedly calls Stelarc a *prophète malheur*, a prophet of doom, it is difficult to see how Virilio's own claims are ultimately any different. As Virginia Madsen points out, '[i]n essence, Stelarc's observations are pure Virilio' (Madsen 1995: 80). Obviously, Virilio is horrified by a future increasingly harnessed to technological speed, yet writes in a manner that has an ideological effect similar to much cyberpunk fiction presenting a technological dystopia that is as fascinating as it is appalling. Overstatement and technological determinism prevent the elaboration of a reflexive relationship to technology; such as has been outlined consistently throughout this book. Instead, Virilio can do little else than simply say 'no' to technology.

Denunciation and the metaphysics of subjectivity

Despite an extensive catalogue of the negative effects of technology extending to over seven books, Virilio has not elaborated on how we might engage differently with technology, beyond simply critiquing its negative aspects. As Conley points out 'Virilio does not see any issue beyond resistance through denunciation' (Conley 1993: 87). On this point, it is instructive to compare Virilio with an earlier French theorist of technology, Jacques Ellul, who constructed a similarly overwhelming portrait of technological domination and advocated resistance through denunciation and heroic self-assertion. Virilio too, seeks resistance through a similarly assertive process. When faced with progressivist ideologies of technology Virilio's simplest strategy is: 'as a Christian, I do the opposite, I say no'; or 'let us not trust' the false freedoms that technology promises (Virilio 1989a: 119). Expanding this position, he has staged the problem of resistance in terms reminiscent of Lyotard's techno-anarchism:

> the question of *freedom* is thus central to the problem of techno-science . . . to what extent can the individual still avoid sensory confusion? To what extent will he be able to keep his distance when faced with the hyper-stimulation of his senses? What new type of dependency or addiction will be produced in the near future.
>
> (Virilio 1995: 119, emphasis added)

On one level, these are important and insightful questions and it is not my intention to question their validity. Instead, I wish to examine the way they are *framed* in terms of individual 'freedom', the ability to avoid technological contamination, and hence, the metaphors of distance and addiction. The question of resistance is clothed in the rhetoric of heroic freedom: the repeated references to Stoic philosophy help to confirm this impression. The question of resistance revolves around a stoical ability to remain uncontaminated, not to 'trust' the ideology of techno-fundamentalism, to say 'no' to further technological mediation. Yet such resistance operates within its own disempowering metaphysic.

To denounce technology by asserting one's own relative freedom, is to fall into a similar trap as that faced by Ellul, in so far as 'freedom remains a metaphysical concept tied to subjectivity and control' (Theile 1995: 216).[3] Placing the question of resistance within such a metaphysic ignores the question of constitutive framing, which on one level Virilio is so intent on exploring in terms of its dispersal. Perhaps this is unavoidable in Virilio's work as a result of the terminology he employs.

Terminology

The problem lies in Virilio's analytical terms such as speed, vectors, and disappearance, all of which implicitly construct a one-dimensional ontology trapped within the confines of the very technological system he opposes. For instance we can question the overemphasis given to the term 'disappearance'. All too often in Virilio's work, the reconstituting capacity of technology is equated with the *disappearance* of something. Virilio himself has stated that 'I have always been interested in missing things . . . people, time, history'. As Kroker perceptively notes:

> Virilio can write *The Aesthetics of Disappearance* because all his texts have focussed on 'absented' subjects: from the absented city of *Bunker Archaeology* and the absented bodies of *Speed and Politics* to the absented (human) vision of *Cinema and War* and *The Sight Machine*.
> (Kroker 1992: 42)

One might well add the absence of the 'human' altogether with the publication of the *Art of the Motor*. But there are serious problems in conflating technological reconstitution with disappearance. Two in particular revolve around the use of a term like disappearance. The first is overstatement: time, space, bodies do not simply disappear, but rather their meanings are *reconstituted*, they still exist, but their meanings unfold through a more abstract framework. Such overstatement also precludes the construction of a space for resistance based around the question of ontological contradiction. In other words, people may not wish for so easy an acquiescence in such disappearance; or they may attempt to reclaim or preserve prior frameworks of meaning outside the sphere of technological abstraction which initiates such disappearance (the current 'backlash' against the Internet may be a case in point). Second, by framing 'disappearance' around the aesthetic, Virilio ignores the existence of several constitutive levels within the social. Social relations with the other can occur at a face-to-face level and at an extended and abstracted level (in Virilio's terms 'tele-proximity'). The second level does not entail the disappearance of the first, indeed it draws its meaning from the prior framework, which structured the significance of intersubjective activity. Yet the twin axes of co-existence and drawing upon a prior framework are elided by Virilio who conflates and aestheticises this process around the question of disappearance.

At this point, one might well consider how to read the epigraph quoted at the beginning of this chapter: '[w]ill Earth become humanity's phantom limb?' Virilio 1995: 127). A reading based around the question of disappearance would lament the severing of humanity from the more earthly grounding which once constrained and supported the meanings of subjective action. While a powerful lament, this doesn't suggest any viable alternative to its own pessimistic conclusion. This would be Virilio's reading, at least as I have argued for it here. A second reading, more in line with this book, might concentrate on the metaphor of the globe and its relation to locality. One can argue that prior modes of being have not simply 'disappeared' under the rule of technology; rather they remain and structure activities that take place on more abstracted, technologically mediated levels. This reading would focus on the ontological importance of the means of perspective.[4] By placing prior modes of constitutive being within the category of the aesthetic nature of their disappearance, Virilio elides the productive role of these 'phantoms'.

A similar observation can be made about the use of 'vector' and 'speed.' Both terms structurally preclude a consideration of the qualitative differences which occur between actions that arise within different constitutive frameworks. For instance, a communicative 'vector', established in a relation of presence of tangibility, is qualitatively different from the movement of information on the Internet, or images transmitted through the vectors of the global media. Yet the term 'vector' tends to collapse these different levels into a single level, through which all vectors pass (as does Lyotard's description of the subject as a 'nodal point'). Similarly, 'speed' can only describe quantitative, but not qualitative difference, thereby eliding questions of how we could create frameworks that would allow a differentiated sense of temporality to unfold. In this way, Virilio's terminology remains complicit, at an analytic level, with that which he opposes.

Perhaps the most efficient way to show how Virilio's work flattens the social out into a single constitutive layer would be to examine a writer who has, in fact, attempted to harness this work in order to engage in a strategy of resistance to the dominant technological framework, something Virilio himself never attempts. Here we examine the work of McKenzie Wark who has attempted to harness Virilio's central concept of the vector so as to strategically read the globalised media society against its own grain.

Harnessing the vector for a radical politics – McKenzie Wark's *Virtual Geography*

Wark's cultural practice involves searching for signs of disruption, excess and contradiction in the postmodern social realm, by which he means the world of global media and other technologies of extension. In *Virtual Geography* (1994), he discusses the way global media powerfully shape our understanding and experience of the world. In particular, global television coverage generates a new kind of subjective experience that brings the world closer but fundamentally alters our relationship to this world. Wark looks at four events – the Gulf War, the collapse of the Berlin Wall, the Tiananmen Square massacre and the Wall Street crash – to demonstrate the particularly postmodern form of disorientation that results when our bearings are established via global media technologies. To a large extent, Wark accepts Baudrillard's (and Virilio's) construction of the social as a world colonised and simulated by media and technology. His phrase, 'virtual geography', denotes this new social realm, where our experience of technologies such as global television reconstitutes our mode of engagement with the world, in the sense that this new geography of experience 'doubles, troubles and generally permeates our firsthand experience of space' (Wark 1994: vii). However, rather than succumb to Baudrillard's resignation, Wark attempts to analyse this new form, in order to locate the 'failures of simulation'. He attempts this by appropriating Virilio's concept of the 'vector'.

Wark uses the vector to describe how new technologies link previously disparate points together. The vector is the trajectory 'along which bodies, information or warheads can potentially pass' (Wark 1994: 11). It is also the result of the 'becoming abstract' of the world, a realm where one can no longer tell where events begin or end, or where information is destined to go. Wark attempts to show how vectors function to displace previous frames of reference. What was once distant is now familiar, a 'perverse intimacy' arises as vectors increasingly invade and displace our prior frame of concrete experiences, primarily through media and communications technology.

As with Virilio, Wark's use of the vector designates an analytical tool that reaches across and bridges together various levels of the social. As with Virilio, Wark claims that technological vectors have hollowed out and thoroughly mediated these prior modes of engagement with the world, until there is nothing outside the vector. His oft-quoted maxim 'we no longer have roots, we have aerials' effectively captures Wark's view of how the globalising process has lifted the subject out from prior frames of engagement. New forms of power are emerging through these vectors. Wark responds to all this by claiming that the struggle for social meaning, the constitution of the self, and the possibility of new forms of community must revolve around forming strategies at the abstract level of the vector. Implicit in this is the belief that meaningful forms of social life can take place entirely at this new level of the social. It is here that one can begin to critique Wark's claims.

Such a critique would commence by noting that any analysis constructed entirely around the concept of the vector, constitutes a social ontology that exists on a single, abstracted level. It would argue that the needs Wark identifies – a need for community and a need to construct an alternative to the present state of commodified social relations – can only be dissolved, rather than fulfilled, by the abstracted social ontology he envisages. The forces for social change might well develop through contradictions that occur between these constitutive levels, rather than on a single level. However, Wark's analysis precludes the recognition of such contradictions. This is so for two reasons. First his use of the vector cannot account for prior levels of social life: as soon as he attempts to analyse them through the vector he has already abstracted them. Second, Wark overstates the degree to which these prior modes of being have been dissolved by the globalised forms of media and communication. However, it remains necessary to see how he would have us respond to a world shaped exclusively by the flow of vectors.

Wark attempts to 'identify forces for social change scattered throughout the vectors' (Wark 1994: xiv). These can be discovered by finding the contradictions that occur within the movement of images along media vectors. Media flows attempt to exclude 'noise' in order to create a narrative logic where the joining of distant points on the virtual terrain can function smoothly. Wark looks for the moments when vectors stray outside this hegemonic path. Because the vector has so thoroughly mediated all prior forms of being, the only response is to engage in a struggle for power through strategic use of the vector itself. We need to think the vector dialectically, for 'immersion in the banality of the vector and its products is the condition for thinking its opposite, for imagining communicative actions' (Wark 1994: 161). How, then, does Wark think the vector dialectically?

He identifies a potential conflict between the forces of capital, which restrict the flow and potential of the vector and those which might disrupt the vector's containment. A project of radical subversion and manipulation, redirecting the vector for one's own purposes could lead to an increase in personal autonomy and new forms of community. Wark frames the struggle thus:

> [i]n its wildest imagination, the vector field imagines itself connecting every point to every other point, the vector field is, potentially a rhizome. Powerful interests prevent it from realising this potential, naturally. By encouraging leakage, spillage, noise, cultural practice can reveal the vector field in potentia. This may involve a departure from liberal notions of the proper time and place for things.
>
> (ibid.)

Here Wark's strategy resembles Derrida's in 'No Apocalypse, Not Now' (1984) where Derrida argues that cultural practices need to speed up in order to track the dynamics of the emerging world order. Featherstone (2000/1) points out how Virilio is himself opposed to such a strategy, suggesting that a 'Museum of Accidents' be established to remind us of the horrors of speed. Indeed Virilio's whole project seems to be about uncoupling humanity from the terror of speed in all dimensions.

What might result from a cultural practice that freed up the rhizomatic potential of the vector? Wark hopes that it might initiate the 'product of a communication system which moves information from beyond the thresholds of discrete communities of interpretation who can bind and limit the free play of meaning, preventing its proliferation' (Wark 1994: 227). We are presented with a mode of subjectivity cast adrift from the social forms which used to support and shape it. Indeed, Wark writes that 'cultural autonomy does not consist of blocking the vector flow, but in seizing upon it as a useful tool of self-definition' (Wark 1994: 162). Hence, we have an emphasis on self-creation, where the self is constituted through the free play of meaning. Such an unconstrained mode of subjectivity can only occur when the subject is severed from its former material ties. As Sharp has remarked, this form of 'autonomy emerges from the reconstruction of the cultural ground to express a mode of realisation of the self that acknowledges no limit' (Sharp 1995: 72). Wark's use of the vector furthers this reconstruction of the cultural ground, since it constitutes the subject in a form that transcends older modes of self-formation, such as embodiment, work, interaction and history. I would want to argue: first, that such transcendence leads to a contradiction in terms of 'autonomy', since the subject drifts in a world of temporary and transient contacts with transparent others, which leads to a crisis in terms of individual ontology; and second, that by severing prior forms of material engagement with the world, one can only further the process of commodification rather than resist it.

We have already considered the problems inherent in embracing such unconstrained subjectivity in other parts of the book. We have seen Heidegger's extensive critique of a self-framing mode of subjecthood under the *Gestell*; and we have examined the contradictions in relation to the Futurist's vision of an autonomous technologised subject. In the next chapter, we trace the limits of such a subject in relation to virtual reality. However, in relation to Wark, I want to use what is perhaps the best example of a site of community based around the flow of abstracted vectors, namely the 'virtual' community constituted in cyberspace, in order to show how the erasure of prior referential frameworks annuls the normative conditions from which a community based around the vector flow might be fully sustained. While I discuss the question of virtual communities in detail in the next chapter, I think that it is instructive to discuss them briefly here within the terms of Wark's theories. First, let us see how Wark understands the struggle for social meanings. He states that:

> [w]ith the spread of the vector into the private realm, a window opens that might be used to create a line along which the communication of intimate and collective feeling might take place. Or it might be used exclusively in the interests of privatised consumption.
>
> (Wark 1994: 19)

Here, Wark tends to misread the terms on which any such choice might be made. We have an example of certain types of experience ('intimate and collective feeling'), which are normally developed though concrete modes of interaction. Wark claims that reconstitution of this experience through the manipulation of a more abstract setting – the vector – is eminently possible. Yet, this claim ignores the fundamental changes that would occur in this experiential content if it were registered via the more abstracted form. Here we can turn to the example of 'virtual communities'.

In his introduction, Wark writes of another site for harnessing the potentials of the vector, that of Rheingold's (1995) 'virtual community'. These 'communities', Wark (1994) writes 'are unanchored in locality, but are made possible by the ever more flexible matrix of media vectors traversing the globe'. Presumably, when Wark writes of communities and social bonds occurring via an empowering appropriation of the vector, he is referring to developments such as these. But the idea that these technologically extended forms of interaction could enable a communication of 'intimate and collective feeling' appears deeply problematic. Lorenzo Simpson has argued that the conditions which enable forms of communication in cyberspace also dissolve the normative conditions from which meaningful social relationships could be carried out. He writes:

> [w]hat happens to the dialectic of social recognition as a result of the representations of telepresence? What of recognition in the new, much-heralded communities carved out of cyberspace, where E-mail's rationalisation of communication is already supplanting face-to-face interaction and the normative features that arise therefrom? In the Hegelian narrative, recognition is supposed to confirm self-certainty. But if what gets recognised by the other is my *constructed* identity, which lacks the vulnerability of my primary identity, then what gets recognised by the other does not confirm me, but only my ideal construction of myself ...
> (Simpson 1995: 159)

This ideal construction, at the expense of more solid and meaningful social ties, is the inevitable result of the ideology of autonomy. Simpson claims that the fleeting and unstable form of such interaction could be met by a whole range of attitudes 'from a suspension of belief to outright cynicism'. He concludes that '[s]ocial interaction in such a space might well lose its validating dimension, at least as we know it' (ibid.). It is this hollowing out of prior forms of the social, from which this dimension emerges, that tends to problematise Wark's hopes for a new community that could create itself from the vector flow. Though he claims the vector has the potential to 'escape the restriction of its abstract potential to the commodity form' (Wark 1994: 177), it seems difficult to sustain. Without the validating dimension Simpson discusses, the meanings of social life can be further commodified. By its nature, the vector cannot allow for the importance of less abstract forms of the social, since it can only constitute intersubjective relations in an extended form. Yet these prior modes of engagement do not disappear, rather they are re-established *through* the commodity form, sold back to us via technologies of extension. Furthermore, if we think about the site where this new autonomy is to occur, we find that the technology that allows for autonomous self-creation through strategic use of the vectorial field works by subordinating subjective experience to the technological framework that mediates it, rather than the reverse. How is it possible to speak of autonomy within such a framework?

Finally, one might want to question the extent to which forces for social change could be located through identifying the contradictions that arise when the flow of vectors stray outside their hegemonic path. We may remember that one of Wark's projects was to identify the 'noise' that emerged from media forms, in order to locate a place for the defamiliarisation of perceptions. Wark is certainly optimistic about the potential of such noise. He writes, echoing the claims of Lyotard, that no matter how 'total, systematic and "posthuman" those relations become, they are subject to crisis, rupture and mutation' (Wark 1995: 13). Here, we can draw a comparison with Virilio's work on the 'accident', in that both privilege the heterogeneous moments that break down an older pattern of rationality. Yet, what can the significance of these moments of 'difference' be, if they occur only on a single level of being? The emergence of the heterogeneous, if it is to make a difference, must necessarily interact with the grounding assumptions that emerge from more concrete forms of social life. By contrast, the possibilities afforded by the disruptive play of the vector occurs in a fundamentally different register because Wark has underplayed the role of the less abstract forms of social life that ultimately ground social and cultural significance. Thus, the disruptive potential he claims is already reified. The promise of emancipation through postmodernity, using Virilio's concepts of the vector and the accident, with an attendant conflation of the various levels of social life, results in an emancipation without meaning.

I have tried to argue here for the importance of understanding the dialectical interplay between different forms of social life, some of which are grounded in stability and in more 'concrete' interactions with the world, while others arise within the more abstracted setting that technology facilitates. Such an approach recognises the importance of relations structured in mutual presence and tangibility, as the basic carriers of social meaning. The danger of technology lies in its ability to transcend these prior forms. This is the danger Virilio so consistently observes and warns against. If such transcendence becomes central, then we will no longer be enthralled, but rather, will have become harnessed to technological 'speed'. Even when not using any particular technology, we will have achieved the Futurist fantasy and become 'motors'. Social life would then be constituted within a single, abstract plane. To conceive of the social in this way, as Lyotard, Virilio, and Wark ultimately all do is to ignore the way different modalities of social integration shape how we think and act in the world and towards each other. The alternative, as I have argued requires a certain cultural reflexivity, recognising the benefits of transcendence while setting limits to the extent of its operation. Neither Virilio, nor Wark, representing the twin poles of overstated pessimism and optimism respectively, seem able to move towards this reflexivity.

7 Psychoanalysis, cyberspace and its discontents
Turkle, Žižek, Brennan

To study 'cyberculture' has become an increasingly complex task. The uses of and discourses surrounding the Internet have multiplied enormously in the last half-decade, making any simplistic claim to judge cyberspace as either good or bad increasingly impossible to undertake. To enter cyberspace has become an increasingly normal activity for many in the West, as more people use the Internet as part of their workplace or leisure activities.

Does this 'banalisation' of the Internet, along with its multiple uses and diverse meanings, mean that there can be no 'grand narratives' made concerning cyberculture? If cyberspace has not radically altered culture and society in the dramatic way some predicted, is it simply the case that cyberspace is merely an extension of everyday life? I want to argue that while cyberculture is by no means a monolithic entity, and that one can find counter-examples to virtually every pronouncement or judgement concerning cyberspace, it remains important to examine the broad cultural framework through which something like cyberspace is understood. In examining cyberspace, we can turn back to the example of the Futurists, and their dramatic claims for the impact of the motorcar or the aeroplane. While the *aesthetic* impact of these objects became rapidly reduced with more widespread use, thus rendering the claims of the Futurists to the category of a historical novelty, it remains the case that both the car and the plane changed society dramatically.

This chapter takes up Robert Markley's suggestion that '[c]yberspace is already marked by competing values about reality and subjectivity, by previous political struggles to naturalise and resist particular constructions of reality' (Markley 1994: 439). In particular, cyberspace seems to be an ideal medium for articulating the anxieties and fantasies of the postmodern subject. Within this context, psychoanalysis provides an ideal point of departure to examine the subject of cyberspace.

Psychoanalysis has always been acutely aware of the dilemmas that underpin any shift from one dominant social formation to another. In *Civilisation and Its Discontents*, Freud argued that the price to be paid for 'civilisation' was a repression of the subject's instinctual drives, an act which led to modern forms of neuroses and related psychic maladies (Freud 1966). Given that Western societies have experienced some dramatic transformations in the past few decades, transformations made possible by the rapid expansion of the techno/information sciences, we can turn to psychoanalysis to ask what is to be gained or lost in such a shift. While psychoanalytic theory has concerned itself largely with the individual subject, it is also true that the relation between individuals and the social formations they inhabit has to a lesser extent preoccupied psychoanalysis. On this point Teresa Brennan has argued that the common preconception of psychoanalysis – that the discipline concerns itself with individuals over against social processes – needs to be corrected. While psychoanalysts have 'observed social dynamics in miniature . . . [much of] psychoanalytic thinking is premised on looking down a telescope the wrong way (Brennan 1997: 213). The theorists considered here, Turkle, Žižek and Brennan, as well as their attendant critics, have all to some extent attempted to reverse the lens and apply psychoanalysis to analyse contemporary social processes.

Most discourses on cyber-technologies revolve around the issue of subject formation, a terrain which similarly concerns psychoanalysis. Psychoanalytic theory also allows us to examine the broader context through which any discussion concerning the politics of the Internet needs to consider. The key question here is to what extent can the subject adopt a radical stance in relation to the Internet based upon a strategy of reflexive performativity, as opposed to being framed by a wider ontological framework that would limit the degree and effect of any performative politics? I examine the former position in the work of Sherry Turkle, the latter in Slavoj Žižek. I then turn to criticisms of Žižek's 'conservative' ontological approach, asking whether it is possible to conduct an ontological critique of cyberspace without reducing the space for a politics of becoming – a criticism levelled at Žižek from a feminist perspective by Verena Conley. To try and answer this key question, I turn to the work of Teresa Brennan, whose work is based in a feminist, psychoanalytic approach which, I argue, provides one way to negotiate between performativity and ontology.

From the mirror to the screen: multiple identities and performative politics

Sherry Turkle's work in *The Second Self* and *Life on the Screen* has explored the way virtual spaces have familiarised what were once regarded as marginal and obscure theoretical claims about the nature of human subjectivity and social interaction. For Turkle, psychoanalysis is doubly relevant for any analysis of contemporary human-technological interaction. First, psychoanalysis has theorised a situation which is becoming increasingly mainstream as humans interact with cyber-technologies and find themselves exploring the ramifications of multiple selves and decentred subjectivities. Second, psychoanalysis' 'imperative to self knowledge becomes newly relevant in the culture of simulation . . . our need for a practical philosophy of self-knowledge has never been greater as we struggle to make meaning from our lives on the screen'

(Turkle 1995: 269). Indeed, for Turkle, cyber-technologies work to confirm the insights of psychoanalysis. Noting that 'Lacan insisted that the ego is an illusion' and that the subject is already split, she goes on to claim that 'computer-science has contributed to this new way of talking: by 'making Gallic abstractions more concrete . . . [I]ts bottom-up, distributed, parallel, and emergent models of mind have replaced top-down informational processing' (Turkle 1995: 178). Cyberspace in a sense actualises more arcane theories concerning multiple subjectivities. The ability to create new identities and engage with absent others on the Internet merely focuses an already-existing phenomenon; 'a zeitgeist of decentered and emergent mind, of multiple subjectivities and postmodern selves' (Turkle: 1995: 124).

Turkle adopts the 'windows' metaphor as a way of understanding this pluralisation process:

> Windows facilitate a way of working with a computer that makes it possible for the machine to place you in several contexts at the same time. As a user, you are attentive to only one of the windows on your screen at any given moment, but in a certain sense, you are a presence in all of them . . . your identity on the computer is the sum of your distributed presence. In practice, windows have become a potent metaphor for thinking about the self as a multiple, distributed system. According to this metaphor, the self is no longer simply playing different roles in different settings, something that people experience when, for example, a woman wakes up as a lover, makes breakfast as a mother, and drives to work as a lawyer. The life practice of windows is of a distributed self that exists in many worlds and plays many roles at the same time.
>
> (Turkle 1997: 1096)

Instead of fragmentation and loss, the Internet allows us to celebrate and 'play' with a disunified self which 'no longer agonises about its disunities' (Turkle 1995: 259). There is no agony because what one does on the net seemingly has no material consequence. Thus the net functions like a kind of flight simulator, allowing us to adjust to the reality of inhabiting a postmodern multiple-identity. Turkle recognises the ambivalent potential of cyber-technologies, but argues that:

> Virtuality need not be a prison. It can be the raft, the ladder, the transitional space, the moratorium that is discarded after reaching greater freedom. We don't have to reject life on the screen, but we don't have to treat it as an alternative life either. We can use it as a space for growth. Having literally written our online personae into existence, we are in a position to be more aware of what we project into everyday life.
>
> (Turkle 1995: 263)

Cyber-technologies then, allow some potential for liberation and/or radicality.

For instance, much has been made of the way in which a character can adopt within a virtual community an identity of the opposite sex. This has the potential to open the subject up to new experiences, such as being inscribed according to a new set of cultural frameworks. As Willson points out 'this may lead to a gendered interactive experience inasmuch as the character may find her/himself encountering gender-based reactions and behaviours from other characters' (Willson 1997: 149). However, Willson points out that to valorise the net for its capacity to create multiple selves is reductionist and ahistorical in so far as the concept of multiple virtual selves 'removes the complexity and depth from the process of self-constitution by limiting perceived influences on this process to singular events/experiences and the possibility of infinite (unrelated) multiplicity' (Willson 1997: 157).

Turkle is aware of how virtual freedom can easily transform itself into a denial of the real. She notes that:

> Instead of solving real problems – both personal and social – many of us appear to be choosing to invest ourselves in unreal places. Women and men tell me that the rooms and mazes on MUDs are safer than city streets, virtual sex is safer than sex anywhere, MUD friendships are more intense than real ones, and when things don't work out you can always leave.
> (Turkle 1995: 244)

Turkle is by no means a simple advocate of the freedoms made possible by the Internet. She recognises, to some extent, that cyberspace is an ambivalent space, and that the subject must negotiate between the polarities of experimentation/radicality and conservative dependency. Yet the terms on which she poses such a choice seem strangely unproblematised. Turkle argues that the Internet can allow the subject, through experimentation with their virtual selves, a degree of reflexivity into the conditions that govern their identity. Yet she pays insufficient attention to how such reflexivity is predicated upon the wider contexts that govern the constitution of subjectivity. On this point Straus argues, 'it is precisely the classical distinction between production (creativity) and simulation (or imitation) which Internet culture tends to blur' (Straus 1997: 99). She goes on to quote from Feldman who writes 'it is not for nothing that theorists like Michel Foucault, Judith Butler and Teresa de Laurentis have politicised the production of subjectivity, tying it to the economic influences and cultural factors that determine it' (Straus 1997: 99). Turkle underplays the fact that postmodern forms of subjectivity are entirely suited to late-capitalist consumer culture – a culture which predicates itself on the construction of scarcity through the promotion of fleeting and unstable forms of identification with the social other through the purchase of commodities and their attendant social meanings.

Straus goes further, and suggests that the kinds of experimentation made possible in cyberspace do not simply free us from a single identity, but instead create the ontological conditions for addiction. She claims that:

> it is possible to normalise fantasy as the preferred experience by structuring our relations to a preferred technology. By blurring the notion of 'self' with 'selves', the Internet makes Freud's healthy narcissism less distinguishable from the pathological kind, and addictions to non-corporeal socialisations normative.
>
> (Straus 1997: 102)

The problems, desires and needs which are structured prior to cyberspace, are not so readily translated to the virtual realm. In the introduction to this book I suggested that specific desires and needs are tied to specific forms of social life. If these forms are merely seen as impediments, to be cast off as quickly as possible, then what happens to the *cultural* meanings that grew and were sustained within such a form? For instance, if I can extend the possibilities of subjective exploration through role-playing on the net, or through reshaping my world in cyberspace, what meaning will this have? Who will recognise *my* creativity through the veil of my virtual selves? The potential for symbolic recognition diminishes when the individual becomes the only reference point for their activities.

The social 'need' to connect, to establish and develop relationships with others has not diminished with the advent of communications technology, as the millions of people using the net can testify. Nevertheless, the type of social connection which occurs is more abstract. This situation is an ambivalent one: it can, of course, lead to people finding it easier to say things on the net that they couldn't say otherwise. For example, a well-known hacker remarks, 'I've found myself saying things on the phone that I couldn't say face-to-face. The net subtracts even the human voice – when you've got nothing, you've got nothing to lose – this could be a breakthrough for humans learning about humans' (Cross 1995: 118). While this sentiment might be welcomed to some limited extent, it remains vital to consider the nature of such a 'breakthrough'. If this radically open context encourages experimentation and plurality, it equally allows for the possibility of withdrawal and disengagement. Social needs, such as communication and co-operation, and individual needs, such as creativity, which require the recognition of an other, are not necessarily fulfilled in a form that dispenses with this other, or allows for a radical ambiguity in terms of their modes of recognition. A Carnegie-Mellon study cited in Hubert Dreyfus' book *On The Internet* (Dreyfus 2001: 4) found that 'Greater use of the Internet was associated with declines in participants' communication with family members in the household, declines in the size of their social circle, and increases in their depression and loneliness' (Kraut *et al.* 1998: 1020).

How might we show that specific needs and desires are tied to a specific social framework? One way is to see the emergence of the ontological contradictions that occur in relation to the Internet. I have already discussed the theoretical possibility of such contradictions, but it is possible to refer to a more concrete manifestation of this phenomenon. The growing rate of Internet addiction, a form of addiction where people remain on-line for many hours a day, and find life increasingly difficult to face outside of the net serves as an example of the contradictions inherent in 'abstract' communication. The rise in counselling of people 'addicted' to the net on US campuses suggests that this phenomenon is becoming a personality disorder in its own right.[1] Yet these people are not simply anti-social. As one computer scientist has remarked, 'these people aren't addicted to playing video games. It wouldn't do the same thing for them. They're *communication addicted* . . . They aren't shunning society. They're actively seeking it' (Curtis, in Rheingold 1995: 151–152). What is it that causes this particular form of communication and society to become addictive? This is where we can return to the question of ontological contradiction.

Ontological contradictions occur when ways of being and acting embedded within a particular constitutive form intersect with a more abstract form. Individual and social needs are often played out differently where the social integration is framed without the presence of the other. For instance, the process of social recognition is radically reconstituted in cyberspace. If our identity is confirmed by the Other's recognition, as in the Hegelian model, the fleeting and unstable form of otherness encountered in cyberspace decentres this process of confirmation. In terms of Internet role-playing, the subjective need for experimentation, play and cultivation of a persona, while appearing to have an ideal forum through which they can take place, may actually founder due to the lack of ground of sufficient social recognition through which experimentation could gain significance. To say this is to recognise that rather than provide empowerment, the net *can* function as an addictive space of desire. Slavoj Žižek has explored precisely these kinds of contradiction using a Lacanian framework to arrive at conclusions very different from those of Turkle.

Žižek – the threat is not the virtual but the real

Žižek argues that cyberspace will in fact produce a radical closure of possibility, rather than a productive openness. He wants to take a 'conservative' position against those like Turkle who celebrate the libratory potential of cyberspace. Indeed, Žižek has taken a counter-position to many of the radical possibilities claimed for cyberspace. Against the sublime openness of cyberspace as a realm for limitless invention and experience Žižek posits cyberspace as initiating an 'unbearable closure of being'. Against the claim that cyberspace enhances creativity through human interactions with technology, Žižek argues that cyberspace in fact promotes a debilitating form of 'interpassivity'. Against the freedoms made possible through virtualised forms of multiple and decentred subjectivities, Žižek claims that 'more often than we think cyberspace is still caught in a hysterical economy' and that the conditions of possibility for virtual identities simultaneously create the conditions for psychosis. The openness made possible through cyberspace does not automatically lead to an expanded realm of possibility; in fact for Žižek, it may lead to 'the paradox of an infinity more suffocating than any actual confinement' (Žižek 1997: 154). The plethora of information does not automatically lead to prosperity; on the contrary, Žižek argues that the conditions of possibility of unlimited contact and access to information may actually lead to a condition of 'informational anorexia' (Žižek 1997: 154).

Consequently, instead of more opportunities for activity, technological choices generate a specific form of passivity. While recognising that new technologies promote a kind of activity – choosing programmes on TV, engaging in debates in virtual communities, Žižek argues that the 'other side' of this phenomenon is 'interpassivity'. Interpassivity occurs when Symbolic order acts in the place of the subject. Technology allows this phenomenon to be more widespread. If interpassivity was once confined to spectacular figures or events, for example Christ who took on our suffering, or plays whose cathartic tragedies relieved us of the need to passively endure our own tragic circumstances, communication technologies allow this relationship to become increasingly central. Thus canned laughter and the manipulation of emotions by advertising relieve the subject of the need to respond. One is reminded here of the *Seinfeld* joke concerning 'phone machine sex'. According to Seinfeld, people are simply becoming too busy to physically engage – the future may well involve leaving sex-messages on the other's answering-machine instead of actual sex. Within an altogether different register one recalls Heidegger's contention that technology allows activity to be replaced by 'functioning' – empty forms of movement and circulation deprived of their meaningful context. In fact Žižek comes close to endorsing something like Heidegger's *Gelassenheit* when he says '[t]he so-called threat of the new media lies in the fact that they deprive us of our passivity, of our authentic passive experience, and thus prepare us for . . . mindless frenetic activity' (Žižek 1997: 122).

For Žižek the important distinction to be made with respect to cyber-environments and their real-world counterparts is not between reality and virtuality, but between appearance and simulacrum. He claims that 'what gets lost in today's digital "plague of simulations" is not the true form non-simulated real, but appearance itself'. The concept of appearance concerns itself with a notion of structural ambiguity, a radical undecidability that resists complete appropriation by the subject. This undecidability is constitutive, it provides the conditions of possibility for meaningful subjective engagement with the world. To explain the notion of appearance Žižek uses the example of a child asking how God's face looks – the answer being that whenever the child encounters a face irradiating benevolence and goodness they have a glimpse of His face. For Žižek:

> the truth of this sentimental platitude is that the Suprasensible (God's face) is discernible as a momentary fleeting appearance . . . it is this dimension of 'appearance' . . . which for a brief moment irradiates the

suprasensible Eternity, that is missing in the logic of the simulacrum. In simulacrum, which becomes indistinguishable from the real, everything is here, so that no other transcendent dimension effectively 'appears' through it.

(Žižek 1998: 1)

The appearance as used by Žižek, resembles what I have argued to be Benjamin's ontological framing of the aura and Heidegger's recasting of vision as 'glancing', in that the appearance allows for a form of reception that is non-totalising, where the subject is not consumed in its mode of engagement with the other.

Thus, what cyberspace threatens is not reality but appearance – conceived as such. Žižek uses the distinction between pornography and seduction to make his point about how appearance is necessarily constitutive. Pornography shows it all – a simulacrum of sex, while seduction works via the play of appearances. Here, one might be tempted to make the same distinction between democracy and e-democracy. Whereas democracy works through the fantasy and promise of something else – the sublime thing – e-democracy is pornographic – every decision laid out for the voter via the practice of instant voting, feedback and access to information. E-democracy is to democractic participation what pornography is to real sex – both are simulacrums whose very transparency undermines the possibility for meaningful engagement. Žižek argues that 'there is no meaning without some dark spot, without some forbidden/impenetrable domain into which we project fantasies which guarantee our horizon of meaning' (Žižek 1997: 160).

For Žižek then, the danger with virtual reality is that it 'carries the phantasmatic logic of social reality to its extreme', thus allowing the subject to fully appropriate the other and trade appearance for the simulacrum. If non-virtual reality was always-already virtualised, Žižek argues that cyberspace nevertheless constitutes a different level of subjective mediation, one which threatens to colonise the non-virtual universe by laying bare the essential fantasy or gap which constitutes subjectivity. This would create an 'unbearable closure of being' – whereby cyber-technologies enact an 'excessive fullness' which fills 'the gap which separates the symbolic surface texture from its underlying fantasy'. In a Virilio-like passage, Žižek describes the closure that may result from the 'death of distance':

> The obverse of this suspension of the distance which separates me from a faraway foreigner is that due to the gradual disappearance of a contact with real bodily others, a neighbour will no longer be a neighbour, since he or she will be progressively replaced by a screen spectre; general availability will induce unbearable claustrophobia; excess of choice will be experienced as the impossibility to choose; universal direct participatory community will exclude all the more forcefully those who are prevented from participating in it. The vision of cyberspace opening up a future of unending

possibilities of limitless change, of new multiple sex organs and so on, conceals its exact opposite: an unheard of imposition of radical closure.

(Žižek 1997: 154)

What exactly produces this radical closure? If the symbolic order or big Other is, according to Lacan, the place where language takes place, where intersubjective communication is possible, this is only because there is an essential undecidability or ambiguity concerning the subject's relation to the symbolic order that is necessary for the subject to function. As Sheila Kunkle notes:

> [t]here is a constitutive lack in the subject when she enters language and the object *a* (the gaze, the voice, the faeces, the breast) falls away; and we can never be certain that shared meanings exist beneath the signifying chain. It is this ambiguity and this lack that both makes a certain relationship to reality possible, and simultaneously creates the conditions for a complete breakdown of the subject's place in the familiar space-time continuum.
>
> (Kunkle 1999: 1)

Žižek argues that cyberspace enacts a radical virtualisation not merely because it creates an artificial phenomenological space, but that the psychic investments that accompany it mean that cyberspace is in fact a symbolic order in the making. He gives the example of a 'rape in cyberspace' something which has acquired canonical status in cyberspace discourse (see Dibell 1993). If one commits a 'rape' in cyberspace, it is according to Žižek, 'closer to impoliteness'. On the other hand it can cause 'emotional catastrophe . . . not reducible to mere words' (Žižek 1997: 140). This middle ground is, Žižek argues, the symbolic order itself. For Žižek, this ambiguity is essential for the function of this 'certain relationship' to reality. He argues that 'in ontological terms: the moment the function of the dark spot which keeps open the space for something for which there is no place in our reality is suspended, we lose our very sense of 'reality' (Žižek 1997: 163). The danger with cyberspace is that it is effectively rewriting the symbolic order in a way which undermines this constitutive ambiguity. The result being that the fantasy frame that sustains subjecthood could collapse leaving the subject mired within the paralysing space of the Real.

In this sense cyberspace merely encapsulates a key direction contained within contemporary capitalist culture; an externalisation of the subject's ego through technological means. James Hurley argues that 'Cyberspace thus presents a heightened version of what Žižek sees as a key tendential shift characterising the logic and lifeworld of postmodernity: the greatly increased handing over of the subject's 'self' to the symbolic order which virtualises/realises that self in the subject's stead' (Hurley 1998: 14). In effect the self and its capacity for action is relocated in a virtualised domain. Žižek argues that cyberspace potentially undermines the grounds through which choice or subjective action could be meaningful, as the virtualised other captures the ego and reduces the space for constitutive fantasy:

> since my cyberspace agent is an external program which acts on my behalf, decides what information I will see and read, and so on, it is easy to imagine the paranoiac possibility of another computer program controlling and directing my agent unbeknownst to me – if this happens, I am, as it were, dominated from within; my own ego is no longer mine.
>
> (Žižek 1997: 142)

In relation to this, it is worth considering how the body is constituted within many constructions of cyberspace. It is something of a mistake to say that many advocates of virtual spaces simply want to 'leave' the body. If we take virtual reality (VR), we can say that it does include a form of embodiment, but one that is profoundly reconstituted. This reconstruction of embodiment allows VR to be the place where corporeality is disavowed.

The reconfigured form of embodiment provides no sense of the subject's body as *situated* in space. Cyber-simulations of embodiment provide no perspective through which the subject could gain a sense of body image, essential according to Merleau-Ponty for the simultaneous generation of subject and environment.[2] This is also the case in Gibson's *Neuromancer* (from which so many developers of VR take their cues), where the virtual subject encounters a distinct lack of embodied perspective. Gibson describes the cyberspace experience in the following manner:

> [b]odiless, we swerve in Chrome's castle of ice we're fast, fast. It feels as if we're surfing the crest of the invading program, hanging ten above the seething glitch systems as they mutate. We're sentient patches of oil swept along corridors of shadow.
>
> (Gibson 1988: 200)

Scott Bukatman points out that at no time in the novel does the virtual subject ever gain a sense of bodily situation: 'the reader of *Neuromancer* is kept in the dark regarding the form of [the] body – the subject of cyberspace never examines *himself*' (Bukatman 1993: 207).

The reconstituted form of embodiment necessarily alters the mode of subjectivity that operates in cyberspace. Without a sense of definable perspective, the boundaries between self and world are irretrievably blurred. The virtual subject encounters no resistance, the body seems not so much engaged with the world, as a passive receptor on which technologically mediated sensations impinge. Cathryn Vasseleu has noted how the sense of tactility is reworked in VR. Tactile sensations are no longer interiorised but are registered on exterior surfaces. In VR some dimensions of tactility are emphasised, others marginalised, regarded as 'passive feedback in the cybernetic compartmentalisation of the faculty of touch' (Vasseleu 1994: 160).

The exteriorisation of embodiment renders a different sense of self and world. The effects of this reconstitution lead to what Virilio calls a 'constitutive dispersal', where perspectival boundaries collapse into 'an open system, in which nobody can find any perceptible, objective limits' (Virilio 1991: 72). The virtual subject becomes a prosthetic subject, its body no longer locatable in time and space since everything merges into a field of external sensations. Hayles writes of the experience of VR that 'when the technologically enhanced body is joined in a sensory feedback loop with the simulacrum that lives in RAM, it is impossible to locate an originary source for experience and sensation' (Hayles 1993: 174).

The fantasy which underwrites virtual forms of subjectivity, is, in this scenario, directly imposed from without. This 'radical decenterment' (Žižek 1997: 142) creates the potential conditions for psychosis. Sheila Kunkle argues that:

> There is a complete lack of consistency and opacity in cyberspace to connect (through the space of a distance) the imagined 'materiality' of one's self to others, for the body itself becomes the real, the horrific void now filled with delusional phantasms of replicant others. In cyberspace, the psychotic's body is not just any type of imagined and fragmented encasement, it is that of a cyber clone; a being who exists in the dimension of the real of body horror in the digital world: a being who is compelled by his replicants, who are in turn the excesses of his disorientated sense of self.
> (Kunkle 1999: 1).

Taking up Žižek's point that the externalisation/virtualisation of the ego enacts a fundamental shift in the co-ordinates from which the subject constitutes itself and engages with the world, Kunkle argues that psychotic forms of delusion, the lack of distinction between self and other, the inability to constitute bodily boundaries, the delusion of oneness are precisely the kinds of experiences available in cyberspace. She concludes that 'the insights of Lacan, Žižek and others, all allow us to see that the machine/computer age is generating its own unique forms of paranoia, that if not inducing psychotic behaviour, at the least brings us to the edge of the abyss' (Kunkle 1999: 6). This claim about the generation of a more generalised form of social psychosis is one that we will turn to when examining Brennan's work.

Traversing the fantasy?

Žižek presents us with a pessimistic scenario, a world whose digitalisation rewrites the symbolic order, undermines the phantastic supplement which gives rise to meaningful action, and results in the death of desire, informational anorexia, and a disabling interpassivity. On the other hand, Žižek admits that we are moving into unchartered territory and writes that 'perhaps radical virtualisation – the fact that the whole of reality will soon be 'digitised', transcribed, redoubled in the Big Other of cyberspace – will somehow redeem real life, opening it up to a new perception' (Žižek 1997: 164). This redemptive possibility has its obvious precursors in Heidegger's equation of the danger/saving power of technology. How would this redemption unfold in psychoanalytic terms? Sheila Kunkle argues that the Lacanian notion of 'traversing the fantasy' might enable the possibility of reflexivity as the subject stands on the brink of the technological abyss. What would underwrite such reflexivity would be 'a certain awareness of the subject as to his "limits" within and beyond the symbolic Law' (Kunkle 1999: 6). Lacan considered this position, and, while noting that it was extreme, nevertheless claimed it was 'one that enables us to grasp that man can adumbrate his situation in a field made up of rediscovered knowledge only if he has previously experienced the limit within which, like desire, he is bound' (Lacan 1981: 276).

There are several problems with such positions. First, like Virilio, Žižek somewhat overstates his claims regarding the fate of ontological closure in the cyberspace era. It is not simply that alternate modes of engagement within cyberspace are possible, but rather that Žižek underplays the degree to which late capitalism as a cultural mechanism can prolong the fantasy structure which is essential to subjectivity. The consumption of mediated images and commodities may well prolong the process of 'acceptance through disavowal' that Žižek elsewhere discerns as fundamental to ideology. However the effect in the Real, the destruction of the environment, the exploitation of the third world, will simply go on.

Žižek's surprisingly reductive position in relation to cyberspace perhaps brings to light a theoretical limitation also evident in his earlier work. In a critique of Žižek's work on ideology, David Holmes argued that Žižek's work, while on the one hand highly sophisticated in relation to the analysis of structures (the dream, the commodity-fetish), was on the other hand somewhat one-dimensional in relation to placing these structures within wider social processes. Holmes argues that Žižek's work suffers from 'the reduction of contemporary social phenomena to psychoanalytic ontologies' (Holmes 1992: 101). In other words, Žižek 'finds enthrallment in placing "Being-in-the-world", ontological questions, before social ontological questions' (Holmes 1992: 111). Hence:

> Žižek does not differentiate between normative frames such as the market, the residues of representational form, or information society as an ontology in its own right, or look at how the subject as agent might be split according to them. Without a theory of social structure, Žižek, by default, capitulates to a reification of the sublime and those 'mechanisms of ideology' on 'one plane'. This criticism would not be justified except for the fact that Žižek tries to speak beyond this plane, to elevate a methodological approach to cultural formation to the status of social ontological claims.
>
> (Holmes 1992: 111)

The lack of differentiation Holmes refers to renders problematic the question of 'traversing the fantasy'. The ontology Žižek presents us with allows little space for reflexivity at a *structural* level, i.e. how one could develop a social theory which might reduce the tendencies for closure, social psychosis, and so on. Indeed the direction of Žižek's work after his major statements on cyberspace in *The Plague of Fantasies* tends to confirm this limitation as he has increasingly emphasised a kind of Lacanian existentialism at the expense of any attempt to ground solutions at the social-structural level. Hurley has also pointed to this shift, noting how Žižek's work is increasingly 'reframing larger structural questions . . . in terms of an individual subject's ethical choice' (Hurley 1998).

The question remains, however, as to whether the problem lies in Žižek's particular construction of cyberspace or whether taking any kind of ontological position is politically ineffective. In what is perhaps the most forthright critique of Žižek's work on cyberspace, Verena Conley suggests the latter. In 'Whither the Virtual' (Conley 1999), Conley objects to what she regards as Žižek's glib dismissal of female 'cyberspace ideologists' in particular his generalised paraphrasing and reductive approach to writers such as Turkle and Allequere Stone. In addition, Conley regards Žižek as representative of a conservative and fundamentally Heideggarian position where 'computers and virtual spaces are seen as contributing to an ultimate mode of enframing and recentring of the subject, be it male, female or of any identity whatsoever' (Conley 1999: 129). While Conley's first point is undoubtedly true in the sense that Žižek doesn't really engage with the work of the 'cyberspace ideologists' he criticises, it is the second objection she raises, concerning Žižek's totalising approach, which will be considered here.

Conley claims that Žižek locates the dilemmas of cyberspace too much within the technology itself. She argues that Žižek's theory 'fails to ask "how computers might help recompose notions of subjectivity in today's conditions"' (Conley 1999: 130). Instead of 'elaborating new models by means of computer-assisted subjectivities, Žižek takes shelter in Lacanian schemas' (Conley 1999: 131). For Conley, Žižek's ontology is too restrictive, too dichotomised to allow for alternative possibilities:

> That there could be other ways of thinking the subject, less as division than as partition, interval, in between – ways that feature prominently in many feminist thinkers, from Luce Irigary to Helen Cixous and many others – never becomes even a remote possibility for Žižek, who remains fixated on a division between being and nothingness.
>
> (Conley 1999: 131)

Conley points out that Žižek's ahistorical understanding of the Symbolic order allows no space for new forms of becoming, it ignores the work of feminists who have attempted to theorise subjectivity beyond the 'Master'. Such work suggests political strategies that can be applied to the new socio-cultural realm made possible through new technologies. However, Žižek, according to Conley, finds shelter and 'ontological safety' in the 'traditional closed universe' (Conley 1999: 133), the Lacanian universe of the Master, the universe of film rather than the computer. Conley claims that a very different strategy is possible rather than Žižek's overly 'defensive' one. Decrying Žižek's conservative reading of Lacan's Symbolic order, she notes that:

> Other theories of language that, instead of a rather rigid symbolic order, emphasise creativity through enunciation, speech acts, the capture of speech, and that may not be 'simply' based on binaries such as being and nothingness, do not figure in Žižek's plan.
>
> (Conley 1999: 133)

Up to a certain point Conley's criticisms strike home. Žižek's Lacanian ontology, especially in his work on cyberspace, is open to a criticism that it is simply too reductive. However her criticism of Žižek's technological determinism needs to be mitigated to some degree by Žižek's comments such as 'it is crucial to maintain open the radical ambiguity of how cyberspace will affect our lives: this does not depend on technology as such but on the mode of its social inscription' (Žižek 1999: 4). It remains true, however, that Žižek never really explores the implications of such remarks.

Conley herself is vague as to the degree to which alternatives to Žižek are realisable. Her own language is, perhaps in opposition to Žižek's 'masterly' pronouncements, hesitant. Hence the virtual is 'unspecifiable', it allows for 'multiple potentials for new connections and unseen relations' (Conley 1999: 134). While Conley tends to come down on the side of a politics of radical performativity as against Žižek's conservative ontology, the question remains as to how effective such a strategy might be. First, as I have said, it remains vague. Second, it gravitates towards a politics of the avant-garde, of the few who can successfully negotiate a politics based in the unstable and unknowable. John Streck has noted 'the crucial flaw of cyberspace is that it elevates the right to speak above all others and all but eliminates the ability to listen' (Streck 1997: 45). Streck's point is that cyberspace allows new forms of becoming only for those who are able to function within its ungrounded domain. Without some kind of ontological approach, the larger contexts whereby relations of identity-formation and intersubjective relations are conducted are downplayed. The potential (of which Conley is keenly aware) for a performative politics to be co-opted within the many choices offered by late-capitalist forms of techno-consumption thus remains. The question as to what would enable a performative politics of the net to go beyond the subjective appropriation of the other still requires consideration. I would answer that one solution lies in a more radical ontological theorising than Žižek's reading of Lacan is able to produce.

Sheila Kunkle notes that Lacan suggested another means of escaping technological domination than the hope that subjects can gain some kind of reflexive capacity as they face the abyss. This other means involves 'a recapturing of the distance between the self and its undead doubles' (Kunkle 1999: 6), so that the externalised virtualised self would be limited in its sphere of operation. Such limitations had, according to Lacan, manifested themselves within prior social forms, the atrophy of which has resulted in an increasing individualism where:

> what we are faced with, to employ the jargon that corresponds to our approaches to man's subjective needs, is the increasing absence of all those saturations of the superego and ego ideal that are realised in all kinds of organic forms in traditional societies, forms that extend from the rituals of everyday intimacy to the periodical festivals in which the community manifests itself.
>
> (Lacan 1997: 26)

These prior forms and rituals constrained individualism and set limits to the projection of the ego. Rather than constrain freedom, they enabled the framework through which a limited individual freedom could acquire significance. Kunkle argues that such rituals form part of the objective space of reality where the symbolic Law structures the ego and allows it 'through narcissism, a hierarchy for its drives', and thus make dialectical communication and things like morality and empathy possible' (Kunkle 1999: 6).

This possibility is more productive and realisable than the alternative, the hope for reflexivity as we are poised on the posthuman threshold. In many ways Teresa Brennan's work takes up the possibilities inherent in Lacan's statement. While not specifically concerned with cyberspace but technology more generally, Brennan's work allows us to understand cyberspace via a critical ontological framework, but one that is historicised and informed by feminist politics. Such an approach allows us to maintain the critical and heuristic benefits of Žižek, but enables us to escape the charges levelled so cogently by Conley.

The age of the ego

For Lacan, the Hegelian dialectic may not necessarily lead to progress, but rather to the Heideggerian-like notion of humankind being enframed by technology. Lacan posits that 'the problem is knowing whether the Master–Slave conflict will find its resolution in the service of the machine' (Lacan 1977: 27). Brennan takes up this point to explore the contemporary 'age of the ego'. Her work explores how 'the technological realisation of a "foundational fantasy" in history constructs and overdetermines the subjective psyche' (Matisons 1998: 26). If Kunkle has elaborated how contemporary society is predisposed towards a certain form of psychosis, Brennan takes this possibility further. For her we are in the midst of a 'social psychosis' whereby the ego makes the world over in its own image. Technology and capital allow the acting out of the ego's projection on a historically unprecedented scale. Thus Lacan's:

theory suggests that the ego can only make the world over in its own image by reducing the lively heterogeneity of living nature and diverse cultural orders to a grey mirror of sameness. And it can only do this by consuming living nature in producing a proliferation of goods and services whose possession become the *sine qua non* of the good life. Of course if nature is endlessly consumed in the pursuit of a totalising course, then that course is dangerous for living; it constitutes a danger to one's own survival, as well as that of others. That, approximately, is the technical, legal definition of psychosis.

(Brennan 1993: 4)

While taking a totalising approach, Brennan's reading of Lacan avoids the claims of an ontological politics based on a reinstatement of the Master, a charge that Conley levels at Žižek. Indeed Brennan is clear on this point stating that:

I am proposing something that makes nostalgic arguments for a patriarchal order beside the point ... For the processes discerned here begin within an omnipotent denial of the mother, a denial that paves the way for the denial of fathers, and that is not to be countered by reinstituting a patriarchal bond cemented by the maternal denial.

(Brennan 1997: 218)

The denial that enacts the foundational fantasy occurs in the shift from the pre- to post-natal stages. Brennan argues that 'an objectifying projection is the condition for subjectivity' (Brennan 1993: 23). The pre-natal stage is subject to an intersubjective economy of energy, what Brennan calls the 'original logic of the flesh' (Brennan 1993: 109). Under this logic, communication is fully interactive, there are two partners whose sense of interconnectedness is absolute, there is no 'separation between thought and substance' (Brennan 1993: 109). The post-natal stage sees the gradual unwinding of this logic. There is a delay for the infant in terms of their gratification. Drawing on the work of Melanie Klein, Brennan argues that this delay is managed by the infant through the hallucination of gratifying objects. These hallucinations mitigate the delayed satisfaction of needs. While this mitigation is never really complete, the substitute controlling fantasy becomes an addictive pleasure in itself (Matisons 1998: 27).

The shift from the pre- to post-natal means a fixing of the spatio-temporal boundaries between self and other, a separation that allows for the development of fantasies of power and domination to be enacted. Whereas the pre-natal intersubjective economy allowed for the free flow of energy between mother and infant, the post-natal economy proceeds to fix energy as the infant inverts a situation of lost control and delayed needs into a fantasy of control over the other and of a world of instant gratification. The foundational fantasy has several qualities: it divorces the mental from the physical, it privileges vision over other senses, it creates a fantasy of individuation and subjective autonomy, and it splits off others into object-like entities which need to be dominated. Arguably, many cyberspace interactions contain at least some of these qualities, as we shall see.

The omnipotent fantasy derived through the ego's will was of course present before the rise of the modern social, but it was not satisfied, or the promise that it could, would and should be satisfied was not 'democratically available'. Technological and capitalist expansion allows the constraints upon this fantasy to be lifted. Brennan traces the social implications where the ego is allowed to operate unfettered. This fantasy begins to expand and gain the upper hand in the 'age of the ego' so that more and more of the world is made over in the ego's image. Brennan draws a parallel between the production of commodities and the control of nature to the foundational fantasy and the infant's relationship to the maternal body – the ego's rigidity has a dialectical counterpart in the things the subject constructs. She notes that:

> we can locate the need to control the environment in an attempt to predict and regulate changes within it, to subject the irregularity of living things to a form of domination in which the ego closes off to itself the truth about itself, by making its dream of fixation come true.
> (Brennan 1993: 43–44)

We can see the paradoxes inherent in this type of 'creativity' in many of the discourses surrounding cyberspace and virtual reality. These discourses link the escape from the material and the natural as a precondition for subjective creativity. Creativity is seen as restricted by present forms of materiality. Physical and social limitations merely restrict the freedom of the subject. As a consequence, it becomes desirable to rid oneself of these limitations as soon as possible. The most obvious manifestation of worldly contamination is the body, which is either jettisoned completely, with Manichean contempt, or radically reconstituted so that its perceptual and sensory fields can be regulated. VR provides the possibility for empowerment, because it is regarded by some as the first step in 'the freeing of life from the chrysalis of matter . . . where information can detach itself from the material matrix and then look back on a cast off mode of being as it arises into a higher dimension' (McKenna cited in Rushkoff 1994: 19–20).

The construction of VR as a redemptive, utopian space inevitably leads to the creation of a hierarchy between virtual and non-virtual subjects and environments. The reductive framing of the physical world as a limit necessarily leads to the wish to transcend it. This sense of a hierarchy is succinctly captured in William Gibson's *Neuromancer* trilogy where characters who inhabit cyberspace are privileged over those who still live in the 'real' world of chaos and poverty, those who still inhabit 'the prison of the flesh' (Gibson 1984: 12).

As the subject–object connections which 'frame human communication and technological constructions' (Brennan 1993: 20) replace the more originary connections based around the mother's body and the natural environment, the manner in which we engage with the world begins to alter. Like Virilio, albeit in a different register, Brennan argues that we are experiencing a kind of 'inertia-effect', that 'while the world appears to get faster and faster, it in fact gets slower and slower' (Brennan 1993: 20). She also claims that the way we understand the world is profoundly altered, that the 'constructed fantasy world' favours modes of 'thought which [emphasise] metaphor and abstraction'. Like Sohn-Rethel who asked 'can there be abstraction other than thought?' (Sohn-Rethel 1978) and Sharp's notion of 'constitutive abstraction', Brennan connects abstract analysis and modes of enquiry with more abstract ways of being in the world. She notes:

> As the fantasmatic coverage of a natural living reality with a socially constructed overlay of as different material order severs connections to the living reality, referents in real felt experience carry less weight in any argument. It is easier to carry a chain of reasoning through on the abstract level than it is to tie it back to the material level, partly because the 'nature' of 'material reality' is confused; it becomes a mixture, of the generative natural chain and the subject-object constructions that overlie it.
> (Brennan 1993: 21)

Brennan is aware of the ambivalence of contemporary theory and political practices which emphasise plurality and discontinuity over against totality and universality. While such theories and practices are aware of the problems inherent in 'grand narratives' their alternatives are also complicit with the fantasised projections of the ego. Thus 'any theory which makes fragmentation and discontinuities a virtue and metaphor a blessing accords with the ego's dynamics' (Brennan 1993: 23). Turkle's rather simplistic celebration of the freedoms of a multiple identity must be considered in this light. We have already seen how much of the discourse surrounding creativity within cyberspace is predicated on a denial of the material. If we take the fantasised projection of the ego, its will to dominate, separate, fragment and homogenise we can see these elements in many cyberspace interactions. We can examine Brennan's work in relation to the contradictions within the concept of virtual communities.

From the heady dreams of the Futurists to the apocalyptic warnings of Virilio, we have traced a shift in the manner in which we are able to integrate socially. The type of relationship first constituted through writing where one engaged with the world across a relation of distance, which was once only a subordinate dimension of society, is now becoming its central mode of existence. For Brennan 'the misinterpretation of Derrida's statement that "there is nothing outside the text" reflects the extent to which the world is becoming nothing more than a text' (Brennan 1993: 24). In Brennan's terms, the ego's age begins when the natural and material limits placed on it are eroded via technology and capital. Contextualised thus, the Internet furthers a transformation that is already taking place.

Yet as social relations in the 'real world' seem increasingly fragmented, claims are made for the Internet to bring about new forms of social connectedness. This points to the ambivalence Brennan locates within the age of the ego – while we have become increasingly closed from each other, we turn to an abstracted form of connection in order to regain a sense of the social whole. The Internet is seen to provide an *open* forum stretching across the globe, where potentially anybody can be allowed to participate. There are many examples of this position. Roy Ascott provides a typical one. 'Interactive telecommunications . . . speaks a language of co-operation, creativity and transformation. It is the technology not of monologue but of conversation. It feeds fecund open-endedness rather than an aesthetics of closure and completion' (Ascot 1991: 115).

Yet this co-operative and conversational potential is not always evident. Even the net's most enthusiastic supporters have noticed that the 'conversation' often takes a very particular form, in that individual performance is often the emphasis, rather than reciprocity. Howard Rheingold, whose work has brought the idea of virtual reality and virtual communities into common parlance, points out how most conversations on the Internet have a theatrical element, often featuring well-known participants thoroughly familiar in networking etiquette, who are greatly outnumbered by silent listeners (Rheingold 1995: 61). This correlates with Brennan's point about the ego's need to assert itself in the world – one could take the positing of statements on the net as yet another example of the production of things in order to simulate a more primordial, but nonetheless denied mode of creativity. Rheingold's observation highlights the manner through which the Internet facilitates a mode of subjecthood based primarily on self-invention. Rather than an openness to the other, much of netspeak is theatrical, based on aestheticising the self, while others who don't fit so easily into this self-active framework remain passive listeners. Given that virtual spaces allow subjective creativity to unfold and be 'enhanced' precisely because the restrictions of otherness are diminished, it is hardly surprising that netspeak is rarely framed by an aura of reciprocity.

In this context, we should register Rheingold's insistence on how fragile virtual communities actually are, how the freedoms the Internet allows often work to *disrupt*, rather than build, communities. He gives the example of a Bulletin Board System, which attempted to form a wider community, but was continually flooded with obscene and aggressive postings from young males, until it was forced to close down (Rheingold 1995: 135). The observation of one user is particularly relevant: she noted that the community was 'choked to death with the consequences of freedom of expression' (Rheingold 1995: 135). Freedom of expression, can in Brennan's terms, simply mean the attempt of the ego to dominate the other and make the world over in its own image.

Rheingold's vision of the virtual community is in many ways typical of contemporary American sensibilities, combining technological utopianism with a pervasive nostalgia. He begins by recognising the way in which social life has been radically restructured. He writes about the disappearance of meaningful public places, which once fostered a sense of community when 'the automobilecentric, suburban, fast-food, shopping mall way of life' took hold. He hopes that cyberspace will be able to become, in turn, a meaningful public site which 'can rebuild the aspects of community that were lost when the malt shop became the mall' (Rheingold 1995: 26). He hopes that local communities will be created on the net, and that these will be able to transcend national frontiers and form 'gateways to a wider realm, the world Net-at-large'. Rheingold's hope for the re-establishment of connection often invokes the nostalgia of the small town, the village square and the malt shop. He hopes that bulletin boards and conference links will provide similarly communal meeting places. He thus recognises the alienation of contemporary life via freeways, fast food, consumerism, whilst refusing to recognise that the virtual realm is all too often underwritten by the same values which led to our 'automobilecentric' way of life. The appeal of the virtual lies in its transcendence, of time, space, presence. Yet the same drive for instant gratification via a technologically-induced transcendence led to the creation of the freeway, the shopping mall, the centrality of fast-food, in the first place.

As Brennan points out:

> [I]nstant gratification 'is no longer a matter of hallucination. It is just around the corner, we will have it in a moment. We can visualise it on the silver screen, in a culture where visuality and virtuality – hallucinatory activity made social – have preeminence' consumer society and service culture are 'markers ... of the extent to which a supposedly individual fantasy is acted out in the social order at large'.
> (Brennan 1993: 217)

Given the abstract condition of a virtual community, the question remains as to how it might develop past its random points of inception. What would ground this community, provide a minimum of stability from which a collective identity could emerge and develop, beyond the insular identification of sameness? In the age of the ego how could such a community accommodate difference? For example, how would a purely virtual community deal with conflict without bypassing it or collapsing completely, as in the example cited earlier. The possibility for disavowing difference or conflict can be contextualised in relation to a wider inability to sustain individual and collective stability. According to Mongardini, contemporary social formations are marked by:

> a greater sense of alienation that makes it increasingly difficult to have relationships that demand more of the personality, such as love and friendship, generosity, forms of identification ... The loss of ability to give meaning to reality is also the product of psychic protection, the desire of

the individual not to put himself at risk by exposing himself to the stimulus of reality he can no longer interpret.

(Mongardini 1992: 62)

The avoidance of risk that Mongardini describes, reinforces Brennan's idea about how the ego avoids anything it cannot appropriate as its own. Such avoidance also undermines the formative role that conflict can play within communities. Traditionally, conflict provides a means of establishing identity and establishes a space for the articulation of differences. As Mouffe points out, 'a healthy democratic process calls for a vibrant clash of political positions and an open conflict of interests' (Mouffe 1993: 6). Yet one wonders about the meaning of any conflict in the virtual community given Brennan's understanding of how the ego projects its own fantasies of creation at the expense of the other. Significantly, Rheingold's idealisation of the virtual community is dominated by *sameness*. He constantly writes of the sense of 'shared consciousness' and the 'experience of groupmind', where there will not only be community but 'spiritual communion' (Rheingold 1995: 110). Once again, behind the positive inflections of groupmind, we can see the ego attempting to homogenise the world in its own image.

Brennan's work provides us with an ontological framework that, at the very least should make us cautious about any claims concerning the empowerment of subjectivity through virtual forms of multiplicity (Turkle) or virtual communities which colonise our relationship to time and space and promote forms of subjectivity that are intolerant of difference. Brennan's provides a Lacanian-derived frame of interpretation; however, unlike Žižek, her frame of analysis is fully historicised and not based narrowly around the law of the father. It thus manages to escape at least some of Conley's concerns.

However, how are we to think of cyberspace in the light of Brennan? She links the psychic dimension (the social psychosis) with the economic and technological dimension. The expansion of the latter allows the projections of the former to expand in scope, which in turn increases the desire for more and more domination of the world. In this sense Bill Gates' claim that e-commerce will help alleviate environmental problems because virtual technologies allow products to be reconfigured in virtual rather than material form (Gates 1996) ignores the manner in which the dominant construction of cyberspace is precisely consumerist in nature where the subject engages with the world through the consumption of images and information. Thus what he calls the 'friction free' nature of exchange will inevitably entrench the dominance of consumerist modes of engagement in then non-virtual environment.

The subjective freedoms and freedom of association, the transcendence of material limitation on the net may allow for a more co-operative dimension to arise, or it may merely go towards extending the age of the ego – the result being that the ego may wish to make over the non-virtual world in the image of the virtual – notwithstanding the environmental and social costs of achieving those freedoms. Brennan's solution, although she does not go into great detail, involves setting structural limits to the ego. This involves 'setting political limits on gain' (Brennan 1993: 192). A return to smaller economies of scale would both redress large-scale exploitation, but also keep the ego in check. This would then allow a space for reflection on the denial of the maternal and a consequent rethinking of the conditions of possibility which underwrite any kind of freedom, expansion or gain. Thus the material realm forms a dialectical relation with the psychic realm. Returning to Lacan's point about historically-prior rituals which help constrain the domain of the ego, we can say that setting limits to gain would necessarily involve a reflexive return to smaller-scale modes of social interaction and economic exchange. These modes would then orient the subject in a way that might enable the more cooperative potential of cyberspace to unfold.

That both Brennan and Žižek point out the possible ontological contradictions inherent in many cyberspace claims concerning freedom, subjective exploration and community does not result in the need to reject these technologies. On the contrary they can provide new and meaningful ways of establishing connection, as both Conley and Turkle have pointed out. The tendency to consider the virtual world and the real world as dichotomous domains is, in part, unavoidable in analysing the cultural constructions of cyberspace. If we are to gain the benefits of the virtual, we need to be able to move reflexively between both. This would mean that the globalised, abstracted and flexible nature of cyberspace encounters needs to be both supported and constrained by forms of sociality prior to and 'outside' the Internet. To recognise this, however is to counter the dominant assumptions of cyberculture. It recognises that social relations constituted through presence and tangibility constitute the basic carriers of social meaning, and not simply impediments to emancipation.

8 Conclusion

The argument presented in this book has proceeded by way of a series of stages. First, I have suggested that a critical theory of technology might most usefully be formulated through considering technology as a broad phenomenon that reconstitutes our modes of knowing and acting in the world. The theorists and cultural movements that appear in the book have been analysed in the light of this approach. Second, I have suggested that the reconstitutive process enabled through technology can best be understood under the rubric of abstraction, considered as much a *material* process as one which occurs in the realm of the conceptual. It is at this point that a crucial issue arises. If technology enables a more constitutively abstract mode of engagement with the world, then to what extent might we wish to welcome or limit this mode? What are the conditions under which we are best able to posit the question?

To ask whether or not a specific technology is 'good' for society, is limited in effectiveness, because it places too much attention on the moral or ethical *content* of a particular technological practice, without understanding how this content is structured and shaped in relation to the social *form* that encompasses it. To ask whether the Internet, for example, is a good or bad technology is less effective in terms of a comprehensive approach, than asking something like: 'To what extent might we wish to embrace a mode of social integration based upon fleeting, intangible relations with others, in a context of radical openness and uncertainty? How do the values which are constructed within this more abstract social modality (of which the Internet forms merely a part) affect our more concrete modes of social integration?' We can think about emerging technologies, such as cloning, in this manner. It would be more important to focus on how cloning cuts across established meanings of the 'human' within an existing constitutive form, than to consider whether cloning technology might be used illegally or for undesirable *ends*.

Each of the chapters has dealt with the question of technology, considered through the work of either individual theorists or broader cultural movements. From an extended analysis of this work on technology we can begin to think about how to engage with the world on this more abstract, technological level. At this point, however, it is possible to identify several broad problems with the approaches considered so far.

The first concerns the politics of *overstatement*. To overstate the degree to which technology has reconstituted our prior modes of being and acting in the world elides the fact that this reconstituting process is itself contradictory.[1] Heidegger's conception of a global enframing at times invites this criticism, as does Virilio's despair at the technological destruction of our perceptual, ontological and environmental habitats. This is not to dismiss the importance of their insights, since they allow an important corrective to the equally deterministic notions of those who posit a technologically utopian future. The problem remains, however, as to how we might envisage a response to technology that could go beyond simply saying 'no', let alone being able to determine to what extent we might welcome the process of technological reconstitution. The problem with overstated accounts of technology goes beyond a mere questioning of the accuracy of their descriptions. More problematic is the fact that such approaches cut us off from a strategy based in reflexivity. This question of reflexivity will be discussed in a moment.

Conversely, there is a problem with attempting to theorise a radical politics of technology entirely within a single abstract level of being. To do this is to ignore the way social formations encompass a variety of constitutive forms, each constituted at a different degree of abstraction. We have seen the more aggressive version of this move, in the case of the Futurists, whose aestheticisation of destruction constructs an abstract mode of engagement, whereby subjective power arises through the transcendence of prior settings and constraints. Even here, however, the residual meanings contained within these settings haunt practices which occur at this more abstract level, as the Futurists attempt to violently reassert the identity of the subject, and the nation, in an abstract setting that would rob both of their autonomy.

Writers such as Lyotard and Wark hope to situate a very different type of politics around the technological, attempting to harness the destabilising and reconstitutive effects of technological mediation for a postmodern politics, based around exploring the possibilities for difference and heterogeneity, as opposed to the modern forms of totalising reason and instrumentality. Proponents of cyberculture look to the technological as a means of both subjective empowerment and the renewal of a sense of community in a kind of technological Eden, valorised precisely because of its ability to transcend the dilemmas of a more concrete form of social life. I have suggested that such attempts are contradictory, in that they fail to recognise how the specific social and cultural meanings, dilemmas or desires that were structured within a prior framework, undergo a shift in register when reconstituted within the terms of a more abstract technological framework. Hence, the claims for liberation, creativity, heterogeneity and difference, which, these writers argue, are more effectively achieved through the embrace of technologies that constitute social life more abstractly, are deeply problematic to the extent that the transcendence of these prior constitutive forms erodes the normative conditions through which the meaning of these terms could unfold.

In contrast to this, I have argued that we engage with the world through a variety of constitutive frames, each working at a different level of abstraction. We have seen throughout the book that technology allows for the transcendence of these prior frameworks. What is the effect of such a transcendence? In answering this, it is crucial to examine how these prior settings are culturally understood. As we have seen, there are a variety of responses to this. One way is to regard these prior frames as mere constraints; symptomatic of a social and cultural malaise (Futurism); to regard such frames as evolutionary cast-offs (the aggressive utopian elements of cyberculture); or to consider them as too bound up with forms of totalitarian reason to be useful to our own era (Lyotard). A second way lies in simply (but powerfully) lamenting the loss of these prior modes of Being-in-the-world (Virilio, and some aspects of Heidegger). Neither approach considers the importance, or even the possibility, of preserving these prior modes of relatedness as constitutive *forms*, which structure and shape social and cultural meanings in their own right. As I have already indicated, the use of the term 'form' indicates a certain phenomenologically and socially constituted relation that provides an ontological and ethical base for behaviour. Those who envisage a strategy for technology based on a conception of social life upon a single, abstract plane, tend to downplay the historical and cultural *embeddedness* of the forms that provide the referential context for our actions within the world. I have suggested that such prior forms provide two crucial elements for determining a critical and ethical approach. First, as I have pointed out, they are productive in that they ground and support the meanings of social and cultural life. Second, such prior forms, based in concrete relations of mutuality and reciprocity, may prove more resistant to the instrumental tendencies of the information age, and the harnessing of desires through the commodity relation, than more abstract forms are able to be.

If this understanding of how social life unfolds through a variety of constitutive frameworks allows us to determine the limits of a politics located entirely at this more abstract level of technologised being, then it also allows us to determine how we might best be able to harness the radical potentials of technology. It is here that I must emphasise the essential *ambivalence* of technology. We can recall Heidegger's one-sided reading of the televised pictures of the earth from the moon. Such pictures can be read both as a symbol of technological domination of the planet, *and* as gesturing towards the need to cultivate more caring relations with our fragile place of dwelling. Similarly, with something like the Internet, it *can* be the place where we can become aware of the existence and needs of others, or it can extend the domination of a subject-centred mode of taking hold of the world, thereby further constituting social relations through the commodity relation. Finally, as Benjamin hoped, technology might be able to initiate a process of mimetic fragmentation, unsettling dominant modes of understanding the world. On the other hand, technology can further entrench our inability to understand or, following Fredric Jameson, cognitively 'map' our place in the world, let alone envisage how to change it.

In saying this I am not, of course, arguing for simply a *way* of approaching technology that might be autonomously cultivated. The reflexive approach I have argued for throughout this book needs to be distinguished from other varieties of reflexivity. The kind of utilitarian approach to reflexivity, argued for by social theorists such as Giddens and Beck, is very different from what is implied within this work.[2] In Anthony Giddens' case, the approach places an undue emphasis upon the autonomous capacity of the social actor to master specific domains, whether it be the use of technique by self-active subject, or the emancipatory potential of the specific expert to intervene and reconstruct specific environments. This approach to reflexivity remains within a single theoretical level in that it fails to consider how these social actions are framed within different constitutive forms. A more effective notion of reflexivity would involve a reflexivity *between* levels, rather than a reflexivity within a single level. Jeffrey Alexander, although he does not refer to specific 'levels', suggests something like this when he writes, contrary to Beck and Giddens, that 'social self-control can emerge only when modernist actors and institutions are embedded in relations of non-reflexive trust and commitment of a decidedly traditional kind' (Alexander 1996: 136).

Alexander's reference to 'relations of non-reflexive trust' and 'traditional' commitment can be expanded through reference to the constitutive levels approach. In analytically separating different levels of social integration and the different degrees of abstraction at which the experience of time, space, the body and subjectivity can be lived, we can more comprehensively analyse the reconstitution of practices and meanings from one level to another. The manner through which a more abstracted level both 'draws upon' and potentially 'thins' out the meanings of practices embedded within a previous level provides the means to a more comprehensive form of reflexivity, in the sense that we are able to theorise the way these levels ought to be held together in relations of uneven dominance. The question concerning technology can only be determined in relation to these intersecting levels.

If the more abstract form of social ontology made possible through technology is marked by a radical openness and multiple senses of possibility, the question remains as to how choices are made within this domain. If, to return to the example of the Internet, it is possible to have extended relations governed by 'webs of personal feeling' *and* at the same time engage within an environment saturated by personal abuse and even a virtual 'rape', then what will determine the choices made? Clearly, to say that such choices are determined on an entirely individualistic basis remains unsatisfactory. In a technologised social setting, governed by a radical degree of openness, where the usual social and cultural settings which help to foster a sense of responsibility are not so readily available, there is a need to develop something more. Indeed, in the context of a powerfully entrenched consumer society, which tirelessly sells us the possibilities made available by a radical degree of choice, we need to be careful about the very grounds on which any choice is made possible. Throughout this book I have suggested that both social meanings and ethical reference points are relatively determined by the social forms which carry them. To argue for a reflexive preservation of historically prior forms is to argue for the maintenance of those reference points which help to determine how we might engage within a more abstract formation. To adopt a Heideggerian idiom, it is a question of cultivating an appropriate sense of 'attunement'. To the extent that social and cultural meanings and values arise within specific constitutive frames, such frames remain significant for two reasons. First, in so far as they function as a binding modality, they provide a semantic resource through which the more groundless and abstract modes of engagement made possible through technology can draw upon. To what extent can we allow the dissolution of such frames and the potential 'crisis' of meaning that will ensue? Second, as a historically and culturally embedded social modality, such frames help to orient any subsequent actions and perceptions that unfold within a more abstract setting. To the extent that any ethics concerns itself with a respect for the other, we might do well to preserve modes of relatedness which, because they are ground in phenomenal co-presence, cause the question of the other to have a certain existential priority.

To argue for the preservation of less abstract modes is not to advocate any kind of regression back to some historically prior form of social relation.[3] Rather it is to point towards the manner in which these prior settings, which function both as semantic resource and an ethical reference point, can help us establish the most productive way of saying 'yes' or 'no' to technology. Throughout this work we have examined the contradictions which occur when social practices are reconstituted at a different level. The contradictory way that abstract relations are lived *as if* they were 'concrete' both grants meaning to the abstract relation and threatens to erode such meaning. As Alison Caddick, drawing upon the work of Sharp, notes:

> where we have no self-reflexive access to those levels of culture that implicitly anchor our humanity, it is exactly the deeply taken-for-granted desires so strongly anchored here that might catapult us beyond those elements of culture which have historically sustained them.
> (Caddick 1997: 69; also Sharp 1985: 72–80)

A reflexive approach must first recognise this contradictory process, in order to understand the importance of maintaining these different levels within the social formation. If we are to reflexively affirm at certain points the importance of more concrete frames of reference over the multiple possibilities that lie within postmodern forms of openness and heterogeneity, this in itself inevitably results in a different way of taking hold of, and living with contradiction. To be sure, any ethical question contains an aspect of choice, the important point in this context being not to resolve the contradiction, but rather make a choice that can best lead to the re-establishment of a co-operative ethic within contemporary life.

To conceive of the social as an intersection of constitutive forms, thus establishes the ontological importance of culturally preserving these historically prior settings which can guide and enrich social life. As technological advances continue at an almost exponential rate, it is time to choose whether to follow a crudely determinist logic, or whether we ought to more carefully consider the implications of discarding a particular mode of being simply because it is historically or technologically 'obsolete'. At a time when intellectual theory has well and truly deconstructed the human essence, and where emerging technologies cause the question of the post-human to gain increasing currency, we find that the ethical question has also risen in prominence. In so far as ethics requires choice, it becomes clear that we cannot simply affirm an ideal of progress or, conversely, succumb to nihilism. As our means of social integration and broad categories of experience are increasingly carried out on a more abstract level, we need to determine the extent to which meaningful and co-operative forms of life can be carried out entirely at this abstract level. Conversely, how might more abstract modes of integration and experience contribute to a more equitable form of social life? Theorising the social formation as consisting of an intersection of these levels provides a guide for determining to what extent individual and social needs can be served at a particular level.

As technology enables our ways of being and acting to be reconstituted at a more abstract level, certain ethical, political, and ontological dilemmas arise simultaneously. Can our actions be ethically orientated within this new setting? Will technological freedoms further draw us into the circuit of the commodity? To what extent does the attempt to resolve human needs technologically work to undermine the grounds through which such resolution could have meaning? These questions have governed my examination of the theorists and cultural movements discussed in this book. They are questions which posit choices that we have never had to make before. I cannot hope to have answered them completely, but if I have allowed them to resonate, then I will have gone some way towards achieving my intention.

Notes

1 Introduction: in the service of the machine?

1 This general position has developed as a composite mix of insights gathered through the theorists considered in this book, but also through reference to the broad position of the *Arena* editorial group whose work on the theorisation of composite layers of social life and the question of intellectual practice has provided a general position from which this specific work on technology takes its starting point. In particular, the work of Geoff Sharp, whose theorisation of the issue of 'constitutive abstraction' in relation to the question of intellectual practice, has been particularly important. See G. Sharp (1985, 1993, 1995). See also J. Hinkson (1987), James and Carkeek (1992), James (1996) and Caddick (1992, 1997) for elaborations of Sharp's thesis within different contexts.
2 To say this does not in any way imply that I wish to aim for 'presence' in the Derridian sense of the word. I would not wish to deny that all forms of engagement are socially constituted, and for this reason alone negate the possibility of such a mode of presence. But to remain at this recognition is to cut oneself off from the crucial distinction between different social ontologies. The distinction that some settings dispense with the need for the actual sense of the other, while others require it as a structural necessity is, I want to argue, vital for an understanding of technology.
3 For a similar description of this more abstract level of social integration from popular culture compare William Gibson's *Neuromancer*: '[s]ummer in the Sprawl, the mall crowds swaying like windblown grass, a field of flesh shot through with sudden eddies of need and gratification' (Gibson 1984: 34).
4 For a more comprehensive discussion of this change in the relation of the individual to society see Sharp (1994) and Caddick (1997).
5 In speaking of 'social form' and its consequential effects upon social and cultural meanings I do not intend a hypostatisation of this term. It is merely a shorthand way of speaking about the specific frameworks of social practices, analytically distinguishable by the degree of abstraction that they are constituted at.
6 Prior levels in this context specifically refers to constitutive levels that are less abstract.
7 Anne Friedberg writes that '[t]he gradual shift into postmodernity is marked ... by the increased centrality of the mobile and virtual gaze as a fundamental feature of everyday life (Friedberg 1993). For a similar claim see Morse (1991).
8 Literally 'letting-be'. See Chapter 2. For Heidegger, human freedom in relation to technology fundamentally involves this process of letting-be.
9 In speaking of poststructuralism and postmodernism in this way, I do not intend to homogenise the work of thinkers as diverse as Derrida, Lyotard, Baudrillard, Haraway and Foucault. I merely wish to speak of them here at a general level

in so far as they help to express the social and cultural transformation that I refer to as postmodernity.
10 The phrase single plane of being comes from Sharp (1985). To speak of theories constituted within a 'single plane' is to refer to the inability to distinguish between the different modes of engagement which constitute our ways of representing and acting in the world.
11 The term ontological contradiction is influenced by Sharp's concept of cultural contradiction (1985: 72). Here I follow Paul James (1996: 28) in using the term ontological contradiction in order to avoid confusion in so far as it might be thought that cultural contradiction refers only to elements of cultural life.

2 Beyond enframing: Heidegger and the question concerning technology

1 As we shall see Benjamin's discussions of aura also focused upon the ability of technology to detach things from their embedded context. While Benjamin was more enthusiastic about the possibilities of this process than Heidegger, I hope to show that any radical potential made possible through such a transformation can only be sustained through the interplay of modes of being, some of which are constituted through the abstracting frame of modern technology, while others *phenomenally* set limits to the extent of any decontextualisation (i.e. the 'face-to-face'). The reading of both Heidegger and Benjamin in this context sees their positions as more similar than many commentators would have us believe.
2 This term is from Phillip Fandozzi (1982). It describes the modern form of nihilism, See *Nihilism and Technology*. The fantasy of infinite possibility is one that frames much of the discourse of cyberculture. See Chapter 7 for an extended critique of technological possibility that extends the perspective outlined in the present chapter.
3 Reasons of space prevent me from dealing more fully with Jünger's work in this book. His work is examined again in relation to Walter Benjamin in the following chapter. One may wish to question why there is not a separate chapter on Jünger in this book as he is used to contextualise both Heidegger's and Benjamin's work at certain points. There are several reasons for this decision. First, the relationship between Heidegger and Jünger has been dealt with in elaborate detail by Zimmerman and there seems little reason to repeat Zimmerman's claims with which I concur. Second, I decided to examine the Italian Futurists instead of devoting a chapter to Jünger (whose work resembles theirs in many ways) because I wished to show that the sensibilities relating to technology as a practical means of abstraction were a more *universal* phenomenon than might be thought if one were to concentrate entirely upon developments within Germany (despite whatever useful insights might result from a more detailed study of one particular historical location). Third, the Futurists convey a more comprehensive sense of a 'technological' sensibility that extends beyond the cult of the machine to all areas of art and life than can be found in Jünger, and thus are more evocative in terms of the constitutive abstraction argument.
4 For a more comprehensive critique of Singer upon the issue of reproductive technologies see Caddick, A., 'Witnessing the Bio-Tech Revolution'. *Arena Journal*, 8, 1997: 59–90.
5 The passage was excised from 'The Question concerning Technology'.
6 Milgram conducted a famous series of experiments where subjects were required to administer a series of electric shocks upon another subject. Milgram noted the connection between the degree of mediation and the willingness of subjects to follow orders and submit the other to the 'shock'. He concluded that 'any force or event that is placed between the subject and the consequences of shocking the victim, will lead to a reduction of strain on the participant and thus

lessen disobedience. In modern society others often stand between us and the final destructive act to which we contribute' (Vetlesen 1993: 374). The recognition that social abstraction can allow the ethical significance of subjective actions to withdraw remains crucial to the argument being made in this book.
7 Ettinger's position is summed up in a chapter of Regis 1990: 145.
8 Thiele (1995) makes this point. This part of my argument is influenced by his reading of what he calls Heidegger's 'disclosive art of politics'.
9 To speak of politics in this way is to distinguish it from the type of political immanentism that Lacoue-Labarthe identifies in Heidegger's political speeches in the thirties. See his *Heidegger, Art and Politics* (1990). Lacoue-Labarthe argues that Heidegger's appeal to an immanent national destiny belies his own non-substantive approach to the philosophy of being. The approach being taken in this chapter reads Heidegger's other statements concerning politics to indicate a constitutive form that would ground and preserve a specific mode of relatedness. One can contrast constitutive forms with their specific contents. Heidegger's political errors can, at least in this context, be understood as pertaining to the latter.

One can also locate such a distinction in Benjamin's work, where following Irving Wolfarth, it is possible to demarcate an authentic form of aura from its false substitutes. In this context, Heidegger's political immanentism would be understand as a false auratic substitute as opposed to politics understood as a mode of revealing vis-à-vis its function as an ontological *form*, which would be an authentic form of aura. See Chapter 3 for more on this.

3 Walter Benjamin and technology: social form and the recovery of aura

1 The term comes from the title of Jeffrey Herf's study, *Reactionary Modernism, Technology, Culture and Politics in Weimar and the Third Reich* (1984).
2 Though this apparently contradictory pair might well be said to constitute the postmodern condition.
3 First, Benjamin notes in the epilogue that the destruction of the aura will result in social destruction if this change does not correspond with the change in property relations. Second, there is Benjamin's understated irony in using Marx's formula for commodity fetishism when describing the destruction of the aura and its impact on the social field.
4 Perhaps it is more accurate to say that the severing of historical experience reconstitutes the desire for transcendence. Within postmodernity, the desire for transcendence has not abated, rather it manifests itself within a hyperindividualised framework (see Chapter 7). Also see Cooper, 'Only a Comet can save us now?' *Arena Journal*, 8, 1997: 1–5, for a discussion of contemporary transcendence in relation to the Heaven's Gate mass suicide.
5 See Chapter 7 for an extended discussion of how virtual reality technologies create a more powerful phantasmagoric effect within postmodernity.
6 Comay compares Benjamin's form of looking with Heidegger's statement that '[e]verything depends on catching sight of [*erblicken*] what comes to presence in *Technik* instead of simply staring [*starren*] at the technological'. See Heidegger (1977: 32).

4 Futurism and the politics of a technological being in the world

1 Lewis wrote that 'Automobilism (Marinettism) bores us. We don't want to go about making a hullo-bulloo about motor cars, any more than about knives or forks, elephants or gas pipe . . .' (Lewis in Hewitt 1993: 206).
2 Jeffrey Schnapp provides a useful summary of this position. See 'Forwarding Address' in *Fascism and Culture*, special issue of *Stanford Italian Review* 8.1–2

(1990): 55–56. For a more sophisticated attempt to separate different political phases of Futurism see S. Falasca-Zamponi, 'The Artist to Power?: Futurism, Fascism and the Avant-Garde', *Theory, Culture and Society*, 13, 2, 1996: 39–58.

3 By 'prior' experience, I do not mean to suggest that the experience need actually to have taken place, nor that it necessarily be empirically prior to reading the text. Rather by 'prior' I wish to suggest the possibility of an experience framed within a social form less abstract than that of the level of the 'text'. In other words 'prior' suggests an ontological, rather than chronological ordering.

4 C.S. Blum (1996) has written a brilliant and extensive study on how gender relations structure what she calls the Futurist 'Fiction of Power'. Her book has been invaluable to my own work on the Futurists. However, while I acknowledge the importance of the way in which gender issues heavily influence the Futurist project, especially in psychoanalytic terms, I have chosen to examine the Futurists more generally in terms of how the question of technology (in the broad sense) structures their work and how their 'Fiction of Power' is made possible through a process of constitutive abstraction.

5 See Chapter 5. More generally see Grosz (1994).

6 Hewitt continues the work of this essay in greater detail in his *Fascist Modernism* (1993).

7 It is interesting to compare the Futurist logic of destruction and liberation with the more contemporary discourse surrounding virtual reality (VR). If the Futurists 'destroy' the body, VR symbolically destroys the body by allowing the subject to inhabit any body at all. As such it destroys the meaning of inhabiting any particular body and the framework through which embodied experience is interpreted. VR also opens itself to the fantasy of pure exchange, as bodies and experiences become instantly interchangeable.

8 Schnapp (1990) argues that what such a logic cannot conceive of is the ethical meaning of loss or destruction. He writes 'at its outer limit, this economy of loss as gain can conceive of the absolute dispersion of an individual. Yet what it *cannot* accept is the tragic: the moment of loss is always made good via the expansion of a larger subject – the collectivity, the army, the nation.

9 Nichols in particular, makes some thematic connections between Marinetti and Baudrillard.

10 This is not of course to suggest that individual violence or national warfare cannot exist in its previous ontological manifestation. I am only suggesting that when both subject and nation are reconstituted within a more abstract framework which problematises the meanings of both – as actions become abstracted from their subjective or national points of origin and individual and national boundaries become obsolete – the reassertion of their identity finds its shortest route through violence in a kind of Hegelian struggle to death without measure.

5 Between totalitarianism and heterogeneity: Lyotard and the postmodern condition

1 This section is influenced through reading James and Carkeek (1992).

2 The other meaning is that the 'postmodern' forms a moment within the modern. Curiously, many supporters of Lyotard tend to emphasise this second meaning to the exclusion of the first.

3 It is not possible within the space of this book to make an extended critique of Lyotard's conception of a postmodern form of justice. Suffice to say his failure to adopt any normative standpoint remains a serious problem. While his theory allows for the making of social arrangements through the notion of the temporary contract, this radically open forum cannot account for the possibility of an already existing inequality between contracting subjects. Indeed his radical pluralism, in

so far as it takes place in the absence of any normative perspective, cannot provide a space where subjects could even reflect upon their political and social circumstances. For a critique of Lyotard along these lines see White (1987/88: 306–319).
4 There are many examples of this sentiment in *Libidinal Economy* (1993), amongst other works.
5 Hinkson's article has been influential in my reading of Lyotard's sublime and the notion of the Information Society as a universalising mode.

6 Paul Virilio: overcoming inertia?

1 See Chapter 7 for more examples of such desires.
2 On this point Slavoj Žižek (1997: 189–190) claims that electronically 'wired' desire is contradictory in that it may lead to the end of desire itself. He writes 'If, in the near future, all data . . . were to become instantly available . . . would not this instant availability suffocate desire?'
3 The comment refers to Ellul, but it could easily apply to Virilio in this context.
4 One could go further, and question the metaphor of the phantom limb and the framing of its disappearance in terms of a privileging of sight – connecting it with questions of technologically reconstituted vision and power. Such a reading of disappearance which I attribute to Virilio would thus contradict a central concern of his in both *War and Cinema* and *The Vision Machine* where vision is detached from a rounded sensory engagement and the eye becomes the central focus for sensory imput and questions of value and judgement.

7 Psychoanalysis, cyberspace and its discontents: Turkle, Žižek, Brennan

1 There are many discussions in computer magazines on this phenomenon, however for consistency I will refer to Rheingold's text, where he discusses the phenomenon of addiction in both the USA and in France. See pages 151–152 and 228–229 of *The Virtual Community* (1995) respectively.
2 See Cooper (1997) for an extended discussion of the cyber-subject. Merleau-Ponty (1962: 133) argues that neither the subject nor the world can simply be posited, but are instead simultaneously generated by the sense of bodily involvement with the external world.

8 Conclusion

1 It is important to once again emphasise that by 'prior' I am referring to social modalities constituted at a lesser degree of abstraction. 'Prior' thus refers to an ontological ordering rather than to a chronological sequence.
2 For an extended discussion of these kinds of reflexivity, see Beck, Giddens and Lash (1994).
3 On the contrary, as Guy Rundle (1989: 95) points out, to advocate that we could simply abandon forms of technological extension is to ignore how these extended forms constitute the 'instruments of moral and theoretical reproduction and dissemination' and thus provide a communicative superstructure that helps prevent moral or cultural regression.

Bibliography

Alexander, J. (1996) 'Critical reflections on reflexive modernization', *Theory, Culture and Society*, 13, 4: 133–138.
Alter, M.P. and Caputo, J.D. (1976) 'Only a god can save us: *Der Spiegel*'s interview with Martin Heidegger', *Philosophy Today*, 20, 4.
Althusser, L. (1971) *Lenin and Philosophy and Other Essays*, London: New Left Books.
Anderson, B. (1991) *Imagined Communities*, London: Verso.
Appolonio, V. (ed.) (1973) *Futurist Manifestos*, New York: Village Press.
Arato, A. (1977) 'The antimonies of a Neo-Marxist theory of culture', *International Journal of Sociology*, 7, 2: 3–24.
Ascot, R. (1991) 'Connectivity: art and interactive communication', *Leonardo*, 24, 2.
Beck, U, Giddens, A, and Lash, S. (1994) *Reflexive Modernisation: Politics, Tradition and Aesthetics in the Modern Social Order*, London: Polity.
Benedikt, M. (1992) *Cyberspace, First Steps,* Cambridge MA: MIT Press.
Benhabib, S. (1984) 'Epistemologies of postmodernism: a rejoinder to Jean-François Lyotard, *New German Critique*, 22: 103–126.
Benjamin, A. (1991) *Art, Mimesis and the Avant-Garde*, London: Routledge.
—— (ed.) (1992) *Judging Lyotard*, London: Routledge.
Benjamin, W. (1968) *Illuminations*, New York: Shocken Books.
—— (1977) *The Origin of German Tragic Drama*, London: New Left Books.
—— (1979) 'Theories of German fascism', *New German Critique*, 17, Spring 1979: 120–128.
—— (1986) *Reflections: Essays, Aphorisms, Autobiographical Writings*, New York: Harcourt Brace.
Bernstein, R. (1991) *The New Constellation: The Ethical-Political Horizons of Modernity/Postmodernity*, Cambridge: Polity Press.
Bloch, E. *et al.* (1982) *Aesthetics and Politics*, London: Verso.
Blum, C. (1996) *The Other Modernism: F.T. Marinetti's Futurist Fiction of Power*, Berkeley: University of California Press.
Blumenburg, H. (1983) *The Legitimacy of the Modern Age*, Cambridge MA: MIT Press.
Bogard, W. (1996) *The Simulation of Surveillance: Hypercontrol in Telematic Societies*, Cambridge: Cambridge University Press.
Bone, P. (1997) 'Ethics: 6 dilemmas', *The Age*, 5/4 (E):1–4.
Brand, S. (1987) *The Media Lab: Inventing the Future at MIT*, New York: Viking.
Brennan, T. (1993) *History after Lacan*, London: Routledge.
—— (1997) 'Social evil', *Social Research*, 64, 2: 210–225.
Buck-Morss, S. (1978) *The Origin of Negative Dialectics*, Cambridge MA: MIT Press.

—— (1992) 'Aesthetic and anaesthetics, Walter Benjamin's Artwork essay reconsidered', *October*, 62: 3–41.
Bukatman, S. (1993) *Terminal Identity*, Durham: Duke University Press.
Caddick, A. (1992) 'Feminism and postmodernism: Donna Haraway's Cyborg', *Arena*, 99/100: 112–128.
—— (1997) 'Witnessing the bio-tech revolution', *Arena Journal*, 8: 59–90.
Comay, R. (1992) 'Framing redemption, aura, origin, technology in Benjamin and Heidegger', in A. Dallery and C. Scott (eds) *Ethics and Danger, Essays on Heidegger and Continental Thought*, Albany: SUNY Press: 139–167.
Conley, V.A. (1993) 'Eco – subjects', in V.A. (Conley (ed.) *Rethinking Technologies*, Minneapolis: University of Minnesota Press: 86–103.
—— (1999) 'Whither the virtual: Slavoj Žižek and cyberfeminism', *Angelaki*, 4, 2: 129–136.
Connerton, P. (1989) *How Societies Remember*, Cambridge: Cambridge University Press.
Cooper, S. (1996) 'Gatesism: the road to nowhere? *Arena Magazine*, 24: 44–49.
—— (1997) 'Plenitude and alienation: the subject of virtual reality', in D. Holmes (ed.) *Virtual Politics: Identity and Community in Cyberspace*, London: Sage.
Crary, J. (1992) *Techniques of the Observer: On Vision and Modernity in the Nineteenth Century*, Cambridge MA: MIT Press.
Cross, R. (1995) 'Modem grrrl', *Wired*, February: 118–120.
Dews, P. (1984) 'The letter and the line: discourse and its other in Lyotard', *Diacritics*, xiv, Fall: 40–49.
Derrida, J. (1984) 'No apocalypse, not now', *Diacritics*, xiv, Summer: 20–31.
Dibbell, J. (1993) 'Rape in cyberspace: a tale of crime and punishment on-line', *The Village Voice*, 21 December.
Dreyfus, H. (2001) *On the Internet*, London: Routledge.
Falasca-Zamponi, S. (1996) 'The artist to power?: futurism, fascism and the avant-garde', in *Theory, Culture and Society*, 13, 2: 39–58.
Fandozzi, P. (1982) *Nihilism and Technology*, Washington: University of California Press.
Featherstone, M. (2000/1) 'Speed and violence: sacrifice in Virilio, Derrida and Girard', *Anthropoetics*, 6, 2: 1–15.
Feenberg, A. (1995) *Alternative Modernity*, Berkeley: University of California Press.
Flint, R.W. (1972) *Let's Murder the Moonshine: Selected Writings of Marinetti*, Los Angeles: Sun & Moon Classics.
Freud, S. (1966) 'Civilization and its discontents', in J. Strachey (ed.) *The Complete Works of the Standard Edition of the Psychological Works of Sigmund Freud*, London: Hogarth.
Friedberg, A. (1993) *Window Shopping: Cinema and the Postmodern*, Berkeley: University of California Press.
Frosh, S. (1991) *Identity Crisis: Modernity, Psychoanalysis and the Self*, London: Macmillan.
Gates, B. (1996) *The Road Ahead*, Harmondsworth: Penguin.
Gibson, William (1984) *Neuromancer*, London: Grafton.
—— (1988) *Burning Chrome*, London: Grafton.
Giddens, A. (1991) *Modernity and Self-Identity: Self and Society in the Late Modern Age*, Cambridge: Polity.
Gill, A. and Riggs, F. (1995) 'The angst of the aura', in *RUATV*, Sydney: Power Institute: 85–103.

Gill, G. (1984) 'Post-Structuralism as ideology', *Arena*, 69: 60–93.
Goslan, R.J. (ed.) (1992) *Fascism Aesthetics, and Culture*, Hanover: University of New England Press.
Grosz, E. (1992) 'Lived spatiality: insect sex/virtual sex', *Agenda*, 26/27: 5–8.
—— (1994) *Volatile Bodies*, Sydney: Allen and Unwin.
Habermas, J. (1983) *Philosophico-Political Profiles*, Cambridge MA: MIT Press.
—— (1991) 'Modernity: an unfinished project', in D. Ingram and J. Ingram (eds) *Critical Theory: The Essential Readings*, New York: Paragon House: 333–344.
Hansen, M. (1987) 'Benjamin, cinema and experience: the blue flower in the land of technology', *New German Critique*, 40:179–225.
Haraway, D. (1985) 'A manifesto for cyborgs', *Socialist Review*, 80: 65–107.
Hayles, N. Katherine (1990) *Chaos Bound*, Ithaca: Cornell University Press.
—— (1993) 'The seductions of cyberspace', in V.A. Conley (ed.) *Rethinking Technologies*, Minneapolis: University of Minnesota Press: 173–190.
Heidegger, M. (1959) *Introduction to Metaphysics*, New Haven: Yale University Press.
—— (1962) *Being and Time*, New York: Harper and Row.
—— (1968) *What is Called Thinking?*, New York: Harper and Row.
—— (1969) *Discourse on Thinking*, New York: Harper and Row.
—— (1971) *Poetry, Language, Thought*, New York: Harper and Row.
—— (1977) *The Question Concerning Technology and Other Essays*, New York: Harper and Row.
—— (1981) 'Only a god can save us: the *Spiegel* interview (1966)', in T. Sheehan (ed.) *Heidegger: The Man and the Thinker*, Chicago: Precedent Publishing: 45–67.
—— (1985) *History of the Concept of Time: Prolegomena*, Bloomington: Indiana University Press.
—— (1993) *Basic Writings*, London: Routledge.
Heim, M. (1993) *The Metaphysics of Virtual Reality*, Oxford: Oxford University Press.
Herf, M. (1984) *Reactionary Modernism, Technology, Culture and Politics in Weimar and the Third Reich*, New York: Cambridge University Press.
Hewitt, A. (1993) *Fascist Modernism*, Stanford: Stanford University Press.
Hillach, A. (1979) 'The aesthetics of politics: Walter Benjamin's "Theories of German fascism"', *New German Critique*, 17: 99–119.
Hinkson, J. (1987) 'Post-Lyotard: a critique of the information society', *Arena* 80: 123–155.
—— (1991) 'Marxism, postmodernism and politics today', *Arena* 94: 138–166.
Holmes, D. (1992) 'Objectifying the sublime', *Arena* (old series) 98: 101–111.
—— (1997) *Virtual Politics: Identity and Community in Cyberspace*, London: Sage.
Honneth, A. (1992) 'Pluralization and recognition: on the self-misunderstanding of postmodern social theorists', *Thesis Eleven*, 31: 24–33.
—— (1993) 'A communicative disclosure of the past: on the relation between anthropology and philosophy of history in Walter Benjamin', *New Formations*, 13: 83–94.
Hurley, J. (1998) 'Real virtuality: Slavoj Žižek and "post-ideological" ideology', *Postmodern Culture*, 9, 1.
Huyssen, A. (1993) 'Fortifying the heart — totally: Ernst Jünger's armoured texts', *New German Critique*, 59: 3–23.
Ingram, D. (1987) 'Legitimacy and the postmodern condition: the political thought of Jean-François Lyotard', *Praxis International* 7: 287–305.
James, P. (1996) *Nation Formation*, London: Sage.
James, P. and Carkeek, F. (1992) 'This abstract body', *Arena* 99/100: 66–85.

Jameson, F. (1992) *Postmodernism or the Cultural Logic of Late Capitalism*, Durham: Duke University Press.

Jay, M. (1992) 'Lyotard and the denigration of vision in twentieth-century thought', *Thesis Eleven*, 31: 34–56.

Jünger, E. (1993a) 'War and photography', *New German Critique*, 59: 24–27.

—— (1993b) 'On danger', *New German Critique*, 59: 27–33.

Kern, S. (1983) *The Culture of Time and Space 1880–1918*, Cambridge MA: Harvard University Press.

Kraut, R. Patterson, M. Lundmark, V. Kiesler, S. Mukophadhyay, T. and Scherlis, W. (1998) 'Internet paradox: a social technology that reduces social involvement and psychological well-being?', *American Psychologist*, 53, 9: 1017–1031.

Kroker, A. (1992) *The Possessed Individual: Technology and Postmodernity*, London: Macmillan.

Kunkle, S. (1999) 'Psychosis in a cyberspace age', *Other Voices*, 1, 3: 1–9.

Lacoue-Labarthe, P. (1990) *Heidegger, Art and Politics: The Fiction of the Political*, Oxford: Blackwell.

Lacan, J. (1977) *Ecrits: A Selection*, New York: Norton.

—— (1981) *The Four Fundamental Concepts of Psychoanalysis*, New York: Norton.

Lash, S. (1999) *Another Modernity: A Different Rationality*, Oxford: Blackwell.

Levin, D.M. (1995) *The Body's Recollection of Being: Phenomenological Psychology and the Deconstruction of Nihilism*, London: Routledge.

Lyotard, J-F. (1971) *Discours/Figure*, Paris: Klincksieck.

—— (1984a) *Driftworks*, New York: Semiotext(e).

—— (1984b) *The Postmodern Condition: A Report On Knowledge*, Minneapolis: University of Minnesota Press.

—— (1985) 'Les Immatériaux', *Art and Text*, 17: 1–16.

—— (1986) 'Response to Kenneth Frampton', *Postmodernism: ICA Documents 4*, London: ICA Publications.

—— (1990) *Duchamp's Trans/Formers*, Venice: The Lapis Press.

—— (1991) *The Inhuman: Reflections on Time*, Oxford: Blackwell.

—— (1993) *Libidinal Economy*, Bloomington: Indiana University Press.

—— (1994) *Lessons on the Analytic of the Sublime*, Stanford CA: Stanford University Press.

McCole, J. (1993) *Walter Benjamin and the Antimonies of Tradition*, Ithaca: Cornell University Press.

Madsen, V. (1995) 'Critical Mass', interview with Paul Virilio, *World Art*, 1: 78–82.

Marinetti, F. (1973) 'Destruction of syntax – imagination without strings – words-in freedom', in U. Appolonio (ed.) *Futurist Manifestos*, New York: Viking Press.

Markley, R. (1994) 'Introduction. Shreds and patches: the morphogenesis of cyberspace', *Configurations*, 3: 1–5.

Matisons, M.R. (1998) 'The new feminist philosophy of the body: Haraway, Butler and Brennan', *The European Journal of Women's Studies*, 5, 1: 9–34.

Merleau-Ponty, M. (1962) *Phenomenology of Perception*, London: Routledge.

Mongardini, C. (1992) 'The ideology of postmodernity', *Theory, Culture and Society*, 9, 2: 55–65.

Morris, M. (1985) 'Postmodernity and Lyotard's sublime', *Art & Text*, 16: 44–67.

Morse M. (1991) 'The Ontologies of Everyday Distraction', in P. Mellencamp (ed.) *Logics of Television*, Bloomington: Indiana University Press: 193–221.

Mouffe, C. (1993) *The Return of the Political*, London: Verso.

Neske, G. and Kettering, E. (1990) *Martin Heidegger and National Socialism. Questions and Answers*, New York: Paragon.
Nichols, P. (1989) 'Futurism, gender, and theories of postmodernity', *Textual Practice*, 3, 3: 202–221.
Norris, C. (1995) *The Contest of Faculties*, London: Methuen.
Nunes, M. (1995) 'Jean Baudrillard in cyberspace: Internet, virtuality and post-modernity', *Style*, 29, 2: 314–327.
Orenstein, P. (1991) 'Get a cyberlife', *Mother Jones*, May/June: 60–64.
Paetzoldt, H. (1977) 'Walter Benjamin's theory of the end of art', *International Journal of Sociology*, 7, 2: 25–75.
Poster, M. (1990) *The Mode of Information*, Cambridge: Polity Press.
—— (1994) 'A second media age?', *Arena Journal* 3: 49–92.
Rajchman, J. (1985) 'The postmodern museum', *Art in America*, October: 111–118.
Readings, B. (1991) *Introducing Lyotard: Art and Politics*, London: Routledge.
Regis, E. (1990) *Great Mambo Chicken and the Transhuman Condition: Science Slightly Over The Edge*, London: Penguin.
Rheingold, H. (1991) *Virtual Reality*, London: Secker and Warburg.
—— (1995) *The Virtual Community: Surfing the Internet*, London: Minerva.
Robins, K. (1995) 'Cyberspace and the world we live in', *Body and Society*, 1: 135–155.
Ronell, A. (1992) 'Support our Tropes', *Yale Journal of Criticism*, 5, 2: 72–76.
Rundle, G. (1989) 'Notes from underground: theory and the green movement', *Arena* 89: 55–69.
Rushkoff, D. (1994) *Cyberia: Life in the Trenches of Cyberspace*. London: HarperCollins.
Schnapp, J. (1990) 'Forwarding address', in *Fascism and Culture*, special issue of *Stanford Italian Review* 8, 1–2: 62.
Schürmann, R. (1987) *Heidegger on Being and Acting: From Principles to Anarchy*, Bloomington: Indiana University Press.
Sharp, G. (1985) 'Constitutive abstraction and social practice', *Arena* (old series) 70: 48–82.
—— (1993) 'Extended forms of the social', *Arena Journal*, 1: 235–236.
—— (1994) 'Keywords: the network', *Arena Magazine*, 10: 46–47.
—— (1995) 'Social form and discourse theory: Foucault and the hidden sociology of post-structuralism', *Arena Journal*, 5: 123–152.
—— (1996) 'The autonomous mass killer', *Arena Journal*, 6: 1–7.
Sim, S. (1996) *Jean-François Lyotard*, London: Prentice Hall.
Simpson, S. (1995) *Technology, Time and the Conversations of Modernity*, London: Routledge.
Sohn-Rethel, A. (1978) *Intellectual and Manual Labour: A Critique of Epistomology*, London, Macmillan.
Straus, N. (1997) 'The fourth blow to narcissism and the Internet', *Literature and Psychology*, 43, 1–2: 96–110.
Streck, J. (1997) 'Usenet and democracy', *New Political Science*, 41/42: 17–46.
Thiele, L. (1995) *Timely Meditations*, Princeton NJ: Princeton University Press.
Turkle, S. (1995) *Life on the Screen: Identity in the Age of the Internet*, New York: Simon & Schuster.
—— (1997) 'Computational technologies and the images of the self', *Social Research* Fall, 64, 3: 1093–1112.
Vasseleu, C. (1994) 'Virtual bodies/virtual worlds', in *Australian Feminist Studies*, 19: 155–168.

Vetlesen, A.J. (1993) 'Why does proximity make a moral difference?: coming to terms with a lesson learnt from the holocaust', *Praxis International*, 12, 4: 371–385.

Virilio, P. (1983) *Pure War*, New York: Semiotext(e).

—— (1989a) *War and Cinema: The Logistics of Perception*, London: Verso.

—— (1989b) 'The last vehicle', in *Looking Back on the End of the World*, New York: Semiotext(e).

—— (1990) *Popular Defence and Ecological Struggles*, New York: Semiotext(e).

—— (1991) *Lost Dimension*, New York: Semiotext(e).

—— (1993) 'The third interval', in Verena Conley (ed.) *Rethinking Technologies*, Minneapolis: University of Minnesota Press.

(1995) *The Art of the Motor*, Minneapolis: University of Minnesota Press.

—— (1996) *Speed and Politics*, New York: Semiotext(e).

—— (2000) *The Information Bomb*, London: Verso.

Virilio, P. and Lotringer, S. (1983) *Pure War*, New York: Semiotext(e).

Wark, M. (1994) *Virtual Geography: Living with Global Media Events*, Bloomington: Indiana University Press.

—— (1995) 'A bugbear is haunting Marxism . . .', *Arena Magazine*, 19: 13.

Weber, S. (1981) 'Walter Benjamin, commodity fetishism, the modern, and the experience of history', in D. Howard (ed.) *The Unknown Dimension*, New York: Basic Books: 245–279.

Wenneberg, B. (1993) 'Ernst Jünger's transformed world', *October*, 62: 43–65.

White D. (1981) 'IVF, towards the industrialization of birth', *Arena* (old series), 58: 23–29.

White, S. (1987) 'Justice and the postmodern problematic', *Praxis International*, 7, 4: 306–319.

Willson, M, (1997) 'Community in the abstract: an ethical and political dilemma?', in D. Holmes (ed.) *Virtual Politics: Identity and Community in Cyberspace*, London: Sage.

Wise, C. (1993) 'The Profane Illumination', *Arena Journal* 2: 195–214.

Wohlfarth, I. (1978) 'No-man's land: on Walter Benjamin's "Destructive Character"', *Diacritics*, 8, June: 47–65.

Wolin, R. (1990) *The Politics of Being*, New York: Columbia University Press.

Zimmerman, M. (1990) *Heidegger's Confrontation with Modernity: Technology, Politics, Art*, Bloomington: Indiana University Press.

Žižek, S. (1997) *The Plague of Fantasies*, London: Verso.

—— (1998) 'Cyberspace, or, how to traverse the fantasy in the age of the retreat of the big other', *Public Culture*, 10, 3. Available HTTP: http://www.uchicago.edu/research/jnl-pub-cult/backissues/pc26/Žižek.html (1 December 2001) Accessed (20 November 2001).

—— (1999) 'The Matrix, the two sides of Perversion', Nettime. Online posting. Available HTTP: http://www.nettime.org/nettime.w3archive/199912/msg00019.html. Posted (2 December 1999) Accessed (20 November 2001).

Index

abstraction 4–5, 21, 33, 38, 42, 71, 77, 80, 91, 103, 160; of the body 92–5; constitutive abstraction 5, 12, 13, 19, 53, 56, 69, 106, 116, 123, 155, 160, 166; Futurist will to 74, 75, 77, 78; literary–cultural process 75; material abstraction 5, 88; sensorial abstraction 57, 60
activity 32, 36; and functioning 25
Adorno, Theodor 49, 54
aestheticisation of technology 47, 49, 50, 51, 52
age of the ego 152–4, 155–6, 157, 158
agribusiness 32, 34, 35
alienation 106
alienated embodiment 127; economic 57; experiential 57; self-alienation 126; social alienation 128, 157–8
allegory 55
ambivalence towards technology 1, 2, 12, 23
antihumanism 46
appearance and simulacrum 144–5
art 30
auratic work of art 59; contemporary art 55; democratic approach to 47; loss of singularity 46; mechanical reproduction 46–7
assembly line production 57
aura 6, 14, 29, 54–5, 62
auratic destruction theory 2, 14, 29, 44, 45–50, 53, 54–5, 56, 73, 89, 90, 168; auratic experience of nature 54; defining 54; function 46; intersubjective realm 65; new forms of 55; primordial notion of 54, 65; and reciprocity 64; reproduction of 63; social structure of auratic experience 63–6; spatio-temporal singularity 46
autonomy 10, 31, 136; cultural autonomy 134; ideology of 135; subjective autonomy 24–5, 154

barbarism 46, 49, 50
Baudrillard, Jean 3, 41, 106, 110
Being 19, 25, 29–30, 31, 34–5, 40; self-showing of the *phainomenon* 34; shared nature of 42; social aspects 42
Being-in-the-world 1, 22, 42, 149, 162
Being-with-Others 6, 41–2
Benjamin, Walter 2, 4, 6, 9, 13, 43, 44–66, 77, 162; Arcades Project 56, 59; auratic destruction theory 2, 14, 29, 44, 45–50, 53, 54–5, 56, 63–6, 73, 89, 90, 168; and the cinema 13, 73; and the phantasmagoria 59–60, 66; and the radical potential of technology 43, 44, 89; and the reciprocity of the gaze 6, 63–5; theories of progress 83; theorisation of technology 45–66, 89
body: abstraction of 92–5; commodification of 81–2; destruction of 81; Futurist body 78–80, 81; liberation from 81; libidinal surfaces 90, 91, 92–5; limitations of the 'natural body' 11; locus of resistance 94; Lyotard on 92–5; meanings of 37–8; organic body 11, 90, 92, 93, 103, 127; of the Other 35, 36; postmodern body 11; reconstitution of 81, 126, 127; renewed focus on 37–8; site of heteronomous desires 92; site of resistance 38; tactile body 37, 38; technological body 79, 81, 117; technological transcendence of 11, 117; Virilio on 126–8, *see also* embodiment
bourgeois culture 49, 75, 76
Brennan, Teresa 15, 139, 152–6, 157, 158–9
business, simultaneity of 73

capitalism 45, 57, 69, 83, 113; abstract logic of modern capital 83; distortion of experience 54; distortion of technology 53;

capitalism – *continued*
  elimination of contingency 112; late capitalism 30, 116, 141, 149; relation of time to 120
chaos theory 99, 100
church 7
city 122
class conflict 77–8
cloning 4, 28, 160
co-presence 7, 12, 13, 42, 95
colonisation, technological 2, 115, 116, 118, 120, 126, 128, 145
commodity exchange 5, 7, 83
commodity fetishism 113, 149
communications technologies 8–9, 14, 16, 42, 74, 122, 125, 142, 144
community 125, 133; virtual communities 134, 135, 141, 143, 144, 156, 157, 158
computers 29, 150
computerisation of society 96, 97, 98, 99; 'windows' metaphor 140
Conley, Verena 150–1, 159
conservation 89
constitutive dispersal 47, 128, 148
constitutive framing 19, 89
consumer society 9
contradiction 31, 41, 43
corporeality 35, 36
cosmos, relationship to 53, 54
creativity 154, 155, 156, 161
cultural despair 113
cultural pessimism 3
cyber-feminism 75, 150
cyber-technologies 11, 89, 139, 140, 145
cyberpunk fiction 1
cybersex 8, 125, 146
cyberspace 3, 4, 15, 29, 75, 118, 135, 138–59, 162; appearance and simulacrum 144–5; consumerist construction 158; interpassivity 143, 144; liberatory potential 143–4; psychoanalytic approaches 2, 138–40, 149, 158; radical virtualisation 146; virtual reality (VR) 13, 73, 82, 108, 145, 147, 148, 154, 156, 169

danger concept 50–1, 52
Dasein 25, 42
death 36
decentralisation 96
dehumanisation 77, 126
dematerialisation 90, 91, 95, 97, 105, 106
democracy: abstracted democracy 41; e-democracy 145

desires 16, 17, 142
destruction 50, 51, 52, 89; aestheticisation of 50, 53, 66, 161; auratic destruction theory 14, 44, 45–50, 53, 54–5, 56, 73, 89, 90
determinism: historical determinism 30, 31; technological determinism 3, 30, 129, 151
difference 96–7, 111, 157, 161
disappearance 115, 128, 130–1
disempowerment 61
distance 21–2, 53
'death' of 145; pollution of 121; shrinking of 47, 48
domination, technological 20, 48, 54, 55, 68, 71, 79, 86, 90, 117, 118, 129, 151
Duchamp, Marcel 108–9

e-commerce 158
ecological destruction 30, 115, 149
electronic capital 122
electronic day 121
Ellul, Jacques 3, 129, 130
embodied understanding 37, 38
embodiment 3, 8, 11–12, 36–7, 38, 79, 85, 93, 94; abstract constitution 81; alienated embodiment 127; contextual process 38; cyber-simulations 147; devaluation of 37; exteriorisation of 148; reconstitution 3, 147; social meanings 4, 11, 92, *see also* body
empowerment 38, 75, 154, 158
enframing 16, 21, 23, 28, 30, 31, 32, 38, 43, 161; *see also Gestell*
essentialism 76–7
euthanasia 118
existentialism 150
experience: abstraction of 52; auratic experience 44, 46, 49, 53, 54, 59, 62, 64, 65, 66, 90; critical stance towards 47; denaturing of 108; distortion of 54; *Erfahrung* 56, 58; *Erlebnis* 56, 58, 61; historical loss of 2, 44, 55–9, 63; impoverishment of 45, 46, 48, 56; ontological categories of 8–12; phantasmagoric forms 45; prior experience 74, 169; subjective 52, 57, 68, 93; universalization of 86

fascism 19, 50, 55, 69, 72, 88, 96
Futurism and 83–4; ritualisation of violence 82
feminine, Futurist denial of the 74, 75
fetishism 63, 64, 73, 80

Index

film and photography 47, 53, 120;
  Benjamin's theory of 73; Futurist
  valorisation of 72–3; and involuntary
  memory 61; 'life as cinema' metaphor 13;
  and mass movements 49; mimetic
  capacity 62, 63; radical potential 13; and
  reciprocity 64; scientific exactitude 61–2;
  and shock 52, 60, 62; technological speed
  72–3
finitude 36, 94
flexibility 95, 101
the future, meaning of 83
Futurism 2, 3, 4, 10, 12, 67–87, 88, 109,
  113, 138, 162; aesthetic theory 75;
  aestheticisation of destruction 50, 53, 66,
  161; attitudes towards women 68, 75; and
  the cinema 13, 72–3; cult of speed 63, 67,
  68, 71–8, 88; denial of the feminine 74,
  75; flying, joy in 71, 79; Futurist body
  78–80, 81; ideology of transcendence 10,
  12–13, 71, 75, 82, 86, 88; logic of flux
  and circulation 88, 113; manifesto 67–8;
  movements within 69; nation-state
  project 16, 84–6, 119; one-dimensional
  ontology 82; power and domination 76,
  79, 83, 86; proto-fascistic elements 69,
  83–4, 86, 117; regeneration project 21,
  71, 72, 73, 88; rejection of inwardness
  and depth 74, 75, 76, 88; 'revolution'
  77–8; Russian Futurism 77; and
  simultaneity 73, 74; subjectivist
  metaphysics 68–9; Tactilism manifesto
  79–80, 81, 82; technological fusion
  fantasy 77; technological imaginary 20,
  118; technological metaphysics 69–71;
  understanding of time 82–3; valorisation
  of technology 12–13, 68; will to
  transcendence 71; worship of the machine
  67, 68, 83, *see also* Marinetti, Emilio

Gelassenheit 14, 25, 29–0, 35, 38, 39, 144
gender difference 94
genetic engineering 96, 116, 118, 126
Gestell 4, 21, 22, 23, 24, 25, 27, 28, 29, 30,
  31, 32, 33, 34, 35, 38, 41, 42, 43, 98,
  134
Giddens, Anthony 14, 163
global 14
grand narratives 91, 95–6, 101, 111, 138,
  155

Heidegger, Martin 2, 3, 4, 6, 9, 13, 18–43,
  47, 48, 57, 67, 161; Being-with-Others 6,
  41–2; critique of autonomous action
30–1; *Dasein* 25, 42; ethical 'indifference'
19, 20, 34; on freedom 20; *Gelassenheit*
('letting be' concept) 14, 25, 29–30, 35,
38, 39, 144; *Gestell* (enframing concept) 4,
21, 22, 23, 24, 25, 27, 28, 29, 30, 31, 32,
33, 34, 35, 38, 41, 42, 43, 98, 134, 161;
historical determinism 30, 31; and
inauthenticity 63; and Nazism 19, 25, 26,
27, 89; nearness and distance 21–2, 23;
and the Other 32–8; and passivity 31, 32;
pessimism 14, 20, 116; political
immanentism 168; preoccupation with
Being 19; on space and time 23; standing-
reserve metaphor 9, 20, 22, 24, 25, 41,
43, 57; and subjectivity 24–5;
technologised subject 16; theorisation of
technology 13, 18–43, 89
heterogeneity 14, 15, 16, 90, 100, 107, 108,
136, 161
heterogeneous fragmentation 115;
  opposition against totalisation 90;
  valorisation 91
historical determinism 30, 31
Holocaust 32, 33, 34, 81
hyperindividualisation 38

idealist progressivism 119
identity: flux of identities 107; identity-
  formation 16, 151; overcoming structured
  identities 75
image-based society 9
in-vitro fertilisation (IVF) 4, 5, 17, 28
inauthenticity 63
incommensurability 111
individual–society relation 10
inertia 116, 120, 155; perceptual and
  sensory 124; temporal inertia 124; Virilio
  on 16, 123–5
information society 7, 91, 95, 96, 98, 103,
  104, 105, 110
information technologies 25, 96, 97, 109,
  121, 123
informational anorexia 144, 148
insecurity 108
instant gratification 157
instrumentalism 3–4, 56, 57, 58, 60–1, 62
Internet 3–4, 17, 131, 140–3, 156, 160,
  162, 163; addiction 3–4, 142–3;
  banalisation of 138; Bulletin Board
  System 156; co-operative and
  conversational potential 156; cyber-
  feminism 75, 150; netspeak 156;
  pluralisation process 140, 141, 142;
  virtual communities 141

180 Index

interpassivity 143, 144, 148
intraorganic dislocation 127

journalism, simultaneity of 73
Jünger, Ernst 3, 26, 27, 45, 50–2, 93, 167
justice 110

Kant, Immanuel 109–10
kinship 6, 10
knowledge 97, 102; embodied knowledge 37

Lacan, Jacques 1, 146, 151–3
language 42, 107; abolition of syntax 67, 74; denial of linguistic materiality 74–5
liberation 15, 20, 24, 84, 86, 91, 152, 161
literature 74; destruction of the 'I' 67, 75–6, 77
local 14, 91, 96
Lyotard, Jean-François 2, 4, 7, 9, 11, 14, 29, 83, 90–113, 123, 161, 162; aesthetic positions 98, 108–11; assessment of scientific theory 99–100; on the body 90, 91; concept of the body 92–5; concept of time 112–13; critique of grand/little narratives 95–6, 101–3; critique of technology 90–1; on Duchamp 108–9; embrace of techno-sciences 106–7, 108; and the information society 98; on knowledge gathering 97; Les Immatériaux 106–8; *Libidinal Ecomomy* 92–5; on performativity 98, 99; postmodern discourse 7, 100–13, 169–70; on subjectivity 103–5; theory of the sublime 109–11

machine: closed and disciplinary machine 109; Duchamp's machines 108–9, 110; love of the 67, 83; metaphor of 83–4; open and emancipatory machine 109
male procreation fantasy 76
manner and matter 35, 38, 39
Marinetti, Emilio 67, 68, 76, 77–8; anti-contemplative aesthetic 74–5; blood metaphor 85; and the cult of speed 71, 72; and the destruction of syntax 74; male procreation fantasy 76; and nationalism 85; and simultaneity 73; and tactilism 79–80, 82; technological metaphysics 69–70
Marx, Karl 5
mass movements 49, 78
masses 47, 48; individualisation of the 78
mechanical reproduction 45, 46–7
media 22, 132, 133, 136

memory 85; collective memory 62; involuntary memory 61; voluntary memory 58
metanarratives 90, 101, 113
militarist ideology 115, 119, 120
military-industrial complex 116, 119
mimetic fragmentation 62, 63, 89, 162
modernity 10; shocks of 52; social settings 10; and temporal experience 62; universalizing narratives 95
mortality 36
mysticism 59, 65

narratives: grand narratives 91, 95–6, 101, 111, 138, 155; legitimation 102; little narratives 101, 102; metanarratives 90, 101, 113; traditional narratives 102
nation-state 7, 16, 69, 88, 119; bodily metaphor 85, 86; Futurist project 84–6; increasing irrelevance 96; permeability 84, 96
nationalism 85
nature: abstraction and reconstitution 76; auratic experience of 54; domination of 54; humanising 54; objectification of 57, 70–1; secrets of 53; technology and 45, 50, 52, 53, 54
Nazism 19, 25, 26, 27, 34, 49, 81, 89
needs, individual and social 16, 25, 142, 143
network society 10
New Age movement 22
new social movements 96
newspapers 57–8, 85
nihilism 23, 24, 26, 27, 37, 49, 82, 89, 115, 165

ontological categories of experience 8–12; *see also* embodiment; space; subjectivity; time
ontological contradiction 16, 23, 25, 31, 43, 77, 82, 84, 143, 167; defining 16
optical unconsious 60, 61–3
organ transplantation 4, 116, 126, 127
Other: concrete Other 35, 36, 37, 41; Heidegger on 19, 32–8; irreducibility of 37–8; ontological status 32; tangibility of 41; universalised 7
overstatement, politics of 128–9, 130, 161

paralogy 14, 101
paranoia 1, 148
passivity 31, 32, 40
pathology of movement 124
patriarchy 69, 88
perception, new models of 120

performance art 116–17
performativity 15, 91, 94, 98, 99, 100, 101, 102, 111, 112, 139, 151
phainesthetics 34
phantasmagoria 59–60, 62, 66
phronesis 39
pluralism 95, 96, 97; of content 97; of forms 97
poetry 30, 35
point of reversal 115, 128
polis: mode of revealing 41; nature of 40–1
pornography 145
possibility explosion 21, 44
post-Marxism 104–5
postcolonialism 96
postmodern body 11
postmodern condition 91, 95
postmodernity 7, 10, 11, 13, 14, 15, 17, 100–13, 136, 169–70
poststructuralism 15
primordial forces 68
proximity and distance 21–2, 23
psychoanalysis 2, 14–15, 138–40, 149

radical potential of technology 43, 44, 45, 49, 66, 89
reality: desacralising 48; inversion 14; objectification of 51
reciprocity of the gaze 63–5
reconstitutive capacity of technology 3, 4, 8, 12, 13, 14, 16, 18, 19, 29, 106–7, 114, 123, 130, 160
reflexivity 2–3, 16, 17, 25, 31, 41, 65, 137, 141, 149, 150, 152, 159, 161, 163, 164
releasement *see Gelassenheit*
reproduction technologies 9, 57–8
revealing 20–1, 33, 40; *alethia* (truth as a revealing) 39, 42; plurality of modes 39–40, 41, 42; politics as a mode of 40–1
Rheingold, Howard 10, 156–7, 158

science, postmodern 29, 99–100
self-creation 134, 136
self-knowledge 139–40
self-objectification 25
shock 45, 51–2, 56, 58, 59, 62; mimetic form of 13, 62, 66, 89; of modernity 52; psychic shock 56
shrinking of the world 21–2
Singer, Peter 28
single plane of being 167
social forms 11, 12, 13, 163–4, 166; constitutive forms 89–90, 162; historical and cultural embeddedness 162, 164; intersecting modalities 12, 13, 164
social integration 5–8, 16, 91, 105, 136–7, 160, 165
social meanings 12, 15, 16, 56, 91
social psychosis 148, 152
social relations 11, 12, 14, 103, 155, 156; agency-extended relations 6, 7, 14, 131; bounded relations 95, 103; as carriers of social meaning 15; disembodied 6, 7, 8, 14; face-to-face 6–7, 8, 42, 56, 65, 85, 95, 102, 131; intersubjective relations 151; ontological modality 7; postmodern social relations 110; primordial relation 42; radically open mode 105; reconstitution of 3, 7, 105; shifting framework 125; technologically mediated 7; temporary contract 105, 112; Virilio on 125–6
space: abstract relations to 9–10, 68–9, 123; colonisation 122, 127; culturally specific spaces 122, 127; destruction of urban space 122; nearness 21–2; technological overcoming of 71; threatened 'habitat' 14, 115; Virilio on 122–3; virtual spaces 10
space-time 122
speed, cult of 63, 67, 68, 71–8, 123; and the cinema 72–3; Futurist valorisation of 71–2; and inertia 16, 116, 125; and regeneration 72; technological speed 72, 115–16, 121, 122, 125, 126, 136; and temporality 120–1; violent speed 72; Virilio on 120–1, 131
standing-reserve metaphor 9, 20, 22, 24, 25, 41, 43, 57, 70
Stelarc 116–17, 118, 129
strategies of resistance 99, 130, 131
struggle and beauty 68
subject formation 10, 11, 84, 139
subject/object dichotomy 58–9
subjectivist metaphysics 24, 68–9, 70, 89
subjectivity 3, 8, 10–11, 24, 134, 141; autonomous subjectivity 38; bourgeois subjectivity 76, 77, 86; decentred subjectivity 58, 76, 77, 99; Heidegger on 24–5; Lyotard on 103–5; multiple subjectivities 140, 141, 155; objectified subjectivity 49; post-structuralist conception of 104; postmodern forms 141; pre- and post-natal 153; proletarian subjectivity 49; reconstitution of 3, 75, 77, 104; subjective activity 25; technological reconstitution 24
subject–object polarity 37
sublime, theory of the 109–11
suicide machine 117–18

182 *Index*

surrogate motherhood 28
surveillance mechanisms 1

tactilism 36, 37–8, 79–80, 81, 82, 147
techno-anarchism 129
techno-fundamentalism 115, 116
techno-sciences 2, 90, 94, 95, 96, 106, 107, 108, 112
technocratic rationality 119
technological determinism 3, 30, 129, 151
technological fusion fantasy 77
technological mediation 2, 3, 13, 16, 18, 48, 69, 119, 128; limiting 2
technological metaphysics 69–71
technology: aestheticisation of 47, 49, 50, 51; danger/saving power of 2, 57, 66, 149; 'essence' of 20, 22, 30, 33; instrumental understanding of 3; radical potential of 43, 44, 45, 49, 66, 89; reconstitutive capacity 3, 4, 8, 12, 13, 14, 16, 18, 19, 29, 106–7, 114, 123, 130, 160; social framing 17; 'unnatural' use of 49, 53; 'yes' and 'no' to 2, 3, 29, 40, 41, 164
television 132
time: abstract relationship to 8–9, 113; colonisation of 2, 120, 127; domestication of 113; Futurist understanding of 82–3; Lyotard on 112–13; naturally grounded notions of 8, 9; prosthetic temporality 124–5; reification of 62–3, 112; simultaneity 9; and speed 120–1; subjective experience of 8, 62; technological measurement 8; technological reconstitution of 71, 120, 121; technological time 121, 122; temporal differentiation 8, 121; threatened 'habitat' 14, 115; Virilio on 8, 120–1
totalisation 14, 15, 98, 99, 100, 119
totalitarianism 15, 16, 96, 123
tourism 14; simultaneity of 73
tradition 46, 50
transcendence 2, 10, 12–13, 43, 58, 71, 75, 82, 86, 88, 118, 157, 162, 168; of the human body 117; human desires for 113; social and cultural 77; subjective 82; technological forms 75, 76, 117, 118, 119

transformative value of work 27
transience 95
transport technologies 22, 74
tribal and peasant societies 6, 7, 8, 9, 10
truth 39, 40, 42
Turkle, Sherry 11, 15, 139–41, 155, 158, 159
twenty-four hour services 9

universalisation 7, 86, 109
utopianism 3, 4, 15, 49, 128, 156–7, 161

vector concept 131, 132–6
victim status 34, 35
Virilio, Paul 2, 3, 4, 7, 8, 9, 14, 23, 47, 114–37, 161; analytical terms 130–1; on the body 126–8; critique of technology 114–17; on inertia 16, 123–5; negative critique of techno-culture 114, 115; one-dimensional critique 14, 15, 115, 116; overstatement 128–9, 130; on scientific exactitude 61; on social relations 125–6; on space 122–3; techno-fundamentalism 115, 116; on technological speed 72, 115–16; technology as reconstituting agent 114; and time 8, 9, 120–1; vector concept 131, 132
virtual communities 134, 135, 141, 143, 144, 156, 157, 158
virtual geography 132
virtual reality (VR) 13, 73, 82, 108, 145, 147, 148, 154, 156, 169
virtual spaces 10, 156

warfare and violence 26, 34, 50, 51, 82, 88, 119–20; aesthetics of destruction 50, 51, 53; as 'cinema' 120; nationalism 85; ritualisation of violence 82; weapons technologies 88
Wark, McKenzie 115, 131, 132–7, 161; and the vector concept 132–6
warrior, technologised 51, 124
will-to-power 89, 116, 117
women, Futurist attitudes towards 75
world as picture metaphor 9, 22, 51, 53
writing 5, 42, 85

Žižek, Slavoj 3, 11, 15, 139, 143–52, 1